Data and Knowledge in a Changing World

CODATA Secretariat

Executive Director: K. Cass
51, Boulevard de Montmorency
75016 Paris, France

E-mail: codata@dial.oleane.com
http://www.codata.org

M. Desaintfuscien

Data Processing in Precise Time and Frequency Applications

With 66 Figures and 15 Tables

ICSU CODATA

Springer

CODATA Secretariat
K. Cass, Exec. Director
51, Boulevard de Montmorency
75016 Paris, France

Author:
Michel Desaintfuscien
21, Domaine de Montvoisin
91400 Gometz la Ville, France

The image on the front cover comes from an animation which shows worldwide Internet traffic. The color and height of the arcs between the countries encode the data-packet counts and destinations, while the "skyscraper" glyphs (or icons) encode total traffic volume at any site. This image was generated by Stephen G. Eick at the AT&T Bell Laboratories.

Library of Congress Control Number: 2006936074

ISBN 978-3-540-48877-4 Springer Berlin Heidelberg New York

Springer is a part of Springer Science+Business Media
springer.com

Typesetting and production: LE-TEX Jelonek, Schmidt & Vöckler GbR, Leipzig
Cover design: eStudio Calamar, Spain

Printed on acid-free paper 52/3100/YL - 5 4 3 2 1 0

to my mother

Introduction to the Series

CODATA is an interdisciplinary Scientific Committee of the International Council for Science (ICSU). The mission of CODATA is to strengthen international science for the benefit of society by promoting improved scientific and technical data management and use. It works to improve the quality, reliability, management and accessibility of data of importance to all fields of science and technology. CODATA is a resource that provides scientists and engineers with access to international data activities for increased awareness, direct cooperation and new knowledge. CODATA, established over 40 years ago by ICSU, promotes and encourages on a world-wide basis the compilation, evaluation and dissemination of reliable numerical data of importance to science and technology. This includes data initiatives and modeling of an interdisciplinary nature such as that encountered in far-reaching projects e.g. Global Change, various Genome projects, environmental and biodiversity issues, etc. Today 23 countries are members, 15 International Scientific Unions have assigned liaison delegates, there are 4 co-opted delegates and 20 supporting organizations from industry, government and academia.

CODATA is concerned with all types of data resulting from experimental measurements, observations and calculations in every field of science and technology, including the physical sciences, biology, geology, astronomy, engineering, environmental science, ecology and others. Particular emphasis is given to data management problems common to different disciplines and to data used outside the field in which they were generated.

CODATA's primary purpose in launching the series "Data and Knowledge in a changing World" is to collect data and the wealth of information pertaining to the intelligent exploitation of data in the conduct of science and technology and to make these data and information available to a multidisciplinary community. This series in support of that goal provides a forum made up of many contributions which can be theoretical treatments, compilations or applied outlines. This includes computer related handling and visualization of data to the major scientific and technical fields.

To this end, the series on Data and Knowledge is open to contributions of various kinds, in particular:

- Fostering the improvement of the accessibility and quality of quantitative and qualitative data;

- Treating classical and ground breaking methods by which numeric and symbolic data are acquired, analyzed and managed;
- Presenting new data and knowledge interfaces designed to optimize interoperability and thereby increase the potential for sharing data among databases and networks;
- Promoting international cooperation in communication and data sharing. This includes works dealing with standardization, data quality agreements and conceptual data descriptions (metadata, syntactic and semantic approaches) along with papers dealing with the evolution of internet based facilities, other forms of worldwide communications and electronic publishing;
- Providing new insights into, or interpretations of, processes leading to creative design in the field of concurrent and/or cooperative engineering, including cognitive aspects critical to data based decision making.

In the evolving information world we live in, where the traditional ways of transferring information as an essential resource are rapidly changing, this Series aims to identify emerging and innovative concepts for data compilation, handling, management, and evaluation. Its ambition is to be a catalyst for change while simultaneously nurturing a thought-provoking forum.

Foreword

The main originality of this book lies in its presentation of an in-depth description of the metrological characterization of very stable frequency sources, such as atomic clocks, as well as the analysis of the principle of their most demanding applications, such as navigation, positioning and very long baseline interferometry.

All these cited fields of interest rely on the measurement of time intervals that necessarily give numerical data: one counts the number of time units occuring between the beginning and the end of an event. Consequently, the analysis of the metrological characterization of stable frequency sources and of the operation of the related scientific and technical applications rely on the treatment of numerical data that can be affected by random and systematic perturbations.

The author presents a rigourous, detailed and unified analysis of the specific signal processing of numerical data arising in that field. Although the subject might seem to be difficult at first sight, the pedagogical talents of Professor Desaintfuscien helps one to easily follow the mathematical derivations and the discussion of their results.

Therefore, this book appears as a reference document for all the scientists and engineers involved in the design or daily use of the related equipment.

Claude Audoin   Paris, January 2007
Directeur de recherche émérite,
Silver Medal (French National Research Center),
Rabi Award (Scientific Committee Annual Frequency Control Symposium)

Contents

1 Introduction 1

Part I Very High Performances Oscillators

2 The Two Parts of an Oscillator 13
2.1 Electrical Resonances 14
2.2 Mechanical Resonances 14
2.3 Atomic Frequency Standards 15
2.3.1 Active Atomic Frequency Standards 16
2.3.2 Passive Atomic Frequency Standard 16
2.3.3 Optically Pumped Cesium Beam 18
2.3.4 Active Hydrogen Maser 27
2.3.5 The Atomic Fountain 30

3 Control of the Local Oscillator Frequency 33
3.1 Active Frequency Standard 33
3.1.1 The Phase Control Loop 34
3.1.2 Gain for the Frequency Noise of the Local Oscillator .. 36
3.1.3 Gain for the Frequency Noise of the Atomic Signal ... 37
3.1.4 Phase Noise of the Output 38
3.1.5 Frequency Error 38
3.1.6 The Best Choices 39
3.2 Passive Frequency Standard 41
3.2.1 Frequency Response for the Frequency Noise of the Local Oscillator 43
3.2.2 Frequency Response for the Frequency Noise of the Atomic Resonance 43
3.2.3 Frequency Response for the Amplitude Noise of the Atomic Resonance 44
3.2.4 Limitation to the Stability Due to the Amplitude Noise of the Atomic Response .. 44
3.2.5 Frequency Error 46
3.2.6 The Best Choices 47

3.3 The Sampled Servo-loop ... 48
3.3.1 Model of the Servo-loop in the Case of an Atomic Fountain ... 48
3.3.2 Frequency Responses for the Frequency Noise of the Local Oscillator ... 49
3.3.3 Frequency Response for the Amplitude Noise of the Atomic Transition ... 51

Part II Characterization of Very High Performances Oscillators

4 Accuracy ... 57
4.1 Definition ... 57
4.2 Estimation of the Uncertainty of a Frequency Source ... 57
4.3 Typical Values ... 60
4.3.1 Primary Frequency Standards ... 60
4.3.2 Commercial Devices ... 61

5 Reproducibility ... 63

6 Stability ... 65
6.1 Definition ... 65
6.2 Measurements in the Time Domain ... 67
6.2.1 The Measurement Process ... 67
6.2.2 Power and Spectral Density ... 75
6.2.3 Variance of the Frequency Fluctuations ... 85
6.2.4 Allan Variance ... 93
6.2.5 Hadamard Variance ... 99
6.2.6 Getting Rid of a Constant Frequency Drift ... 107
6.2.7 The Best Variance for a Given Application ... 112
6.2.8 Practical Time Domain Measurements of Frequency Stability ... 121
6.2.9 Typical Values ... 122

Part III Applications

7 Time and Frequency Metrology ... 129
7.1 Time Scales ... 130
7.1.1 The Main Time Scales in Use ... 131
7.1.2 Algorithms for the Generation of Time Scales ... 137
7.2 Comparison of Clocks ... 142
7.2.1 Clock Synchronization in a Rotating Frame ... 142
7.2.2 One-way GPS Measurements ... 143

7.2.3 GPS Common-view . . . 145
7.2.4 GPS Carrier-phase Time Transfer . . . 146
7.2.5 Two-way Satellite Time and Frequency Transfer (TWSTFT) . . . 147

8 Global Positioning System . . . 151
8.1 The GPS Infrastructure . . . 152
8.1.1 The Space Segment . . . 152
8.1.2 The Control Segment . . . 153
8.1.3 The User Segment . . . 154
8.2 GPS Data Processing . . . 154
8.2.1 Data Transmitted by the Satellites . . . 154
8.2.2 Data Processing by the Receiver . . . 161
8.2.3 Other Algorithms . . . 167
8.2.4 GPS Augmentation: WAAS and LAAS . . . 171
8.3 Other Global Systems . . . 172
8.3.1 GLONASS – GLObal'naya NAvigatsionnaya Sputnikovaya Sistema . 172
8.3.2 Galileo . . . 173

9 Very Long Base Interferometry . . . 175
9.1 Principle of VLBI . . . 176
9.1.1 Interferometry . . . 177
9.1.2 Processing of the Signals . . . 185
9.2 Applications of VLBI . . . 187
9.2.1 Astronomy . . . 187
9.2.2 Geodesy . . . 187

Part IV Appendix

A Useful Integrals . . . 191
A.1 Calculation of Variances . . . 191
A.2 Calculation of Allan Variances . . . 192
A.3 Calculation of Hadamard Variances . . . 192
A.4 Calculation of the Three-samples Variance . . . 194

B Some Calculational Details . . . 195
B.1 The Allan Filter . . . 195
B.2 The Hadamard Filter . . . 195
B.3 Three-samples Filter . . . 197
B.4 Phase Noise Versus Frequency Noise Spectral Density . . . 197
B.5 Phase Noise . . . 198
B.6 Sampling and Hold . . . 199
B.7 Algorithm for a GPS Receiver . . . 200

B.7.1 The Data 200
B.7.2 Algorithm 201

References 205

Index 215

List of Figures

2.1 Block-diagram of an oscillator 13
2.2 Robinson oscillator 14
2.3 Pierce oscillator 15
2.4 Block diagram of an active hydrogen maser 17
2.5 The two parts of a cesium beam frequency standard 17
2.6 Block diagram of a cesium beam frequency standard 19
2.7 Transition probability of an atom across a Ramsay cavity 22
2.8 Central part of the pattern of Fig. 2.7 22
2.9 Mean transition probability of a beam with a velocity distribution 23
2.10 Central part of the curve of Fig. 2.9 24
2.11 Generation of the frequency f_1 25
2.12 Another electronic device generating the frequency f_1 25
2.13 The excitation of the Ramsay cavity as the output of a DRO [90] 26
2.14 Some frequencies generated by intermediate oscillators [112] .. 26
2.15 Production of the maser effect 29
2.16 Phase locking of the local oscillator on the atomic emission 30
2.17 Schematic diagram of an atomic fountain 30
2.18 Calculated Ramsay fringes in an atomic fountain 32

3.1 Phase lock loop for the different components of the signals 35
3.2 The phase lock loop with the phase as the quantity of interest 35
3.3 Equivalent set-up for the local oscillator noise 36
3.4 Equivalent set-up for the noise of the atomic signal 37
3.5 The loop for the nominal frequency of the local oscillator and of the atomic oscillator 39
3.6 Control of the frequency of the local oscillator in a passive atomic frequency standard 42
3.7 Model for the noise in the frequency control loop 42
3.8 The loop for the nominal frequency of the local oscillator and of the atomic transition 46
3.9 Sampling in the loop of Fig. 3.7 49
3.10 The sampled loop for the local oscillator frequency noise 50
3.11 The sampled loop for the amplitude noise 51

3.12 The sampled loop for the frequency noise of the atomic transition ... 52

6.1 Squared amplitude response of the filter versus the reduced frequency ... 68
6.2 Squared amplitude response of the filter associated to the frequency measurement with no dead time ... 70
6.3 Frequency measurement with dead time ... 70
6.4 Squared amplitude response of the filter associated to the frequency measurement with a dead time ... 71
6.5 Squared amplitude response of the filter associated to the frequency measurement with no dead time and association of three successive results ... 72
6.6 Spectrum of the discrete-time process compared to that of the continuous-time process ... 73
6.7 Aliasing when frequency is measured without a dead time ... 74
6.8 Squared amplitude response of the Allan filter ... 95
6.9 Squared amplitude response of the Hadamard filter for $M = 4$ 102
6.10 Squared amplitude response of the modified Hadamard filter for $M = 8$ ... 102
6.11 Squared amplitude response of the balanced Hadamard filter with $(M = 4)$ ... 109
6.12 Squared amplitude response of the three-samples filter ... 110
6.13 Model of frequency noise showing the five power laws ... 114
6.14 Comparison of Allan (*lower curve*), Hadamard and three-samples (*upper curve*) variances ... 115
6.15 Allan variance with $h_1 |\nu|$ noise (*lower curve*) and with $h_2\nu^2$ noise (*upper curve*) ... 116
6.16 Allan variance and the two asymptotes corresponding to h_1 and h_2 ... 117
6.17 Ratio of the coefficients of the three branches for three variances 118
6.18 Three-samples variance. The width of the frequency domain where the component $h_{-1} \times |\nu|^{-1}$ of the frequency noise is dominant is three decades ... 118
6.19 Allan variance. The width of the frequency domain where the component $h_{-1} \times |\nu|^{-1}$ of the frequency noise is dominant is three decades ... 119
6.20 Allan variance with a constant frequency drift ... 120
6.21 Hadamard variance with a constant frequency drift ... 121
6.22 Frequency measurement in the case where a reference oscillator can be used ... 122
6.23 Stability measurement with two identical oscillators ... 122
6.24 Frequency noise spectral density of the oscillator of [72] ... 124

7.1 A GPS satellite and two receivers on the Earth. The distance Earth–satellite is about four times the Earth's radius 145
7.2 The principle of the two-way technique 147
7.3 Paths of TWSTFT 148

8.1 Orbits of the GPS satellites 152
8.2 L_1 and L2 signals emitted by the GPS satellites 155
8.3 Modulo 2 (XOR) addition of the navigation message and the C/A-code and the P-code 156
8.4 Easy retrieval of the navigation message 157
8.5 Measurement of the raw pseudo-range (modulo 299 792.458 m) 162
8.6 Good GDOP 165
8.7 Poor GDOP 166

9.1 The principle of VLBI 175
9.2 The principle of interferometric measurements 177
9.3 Schematic block diagram of a cross-correlator 187

List of Tables

3.1 Phase noise of a commercial quartz oscillator 41
3.2 Guaranteed values of the Allan deviation of a commercially available cesium frequency standard 45

4.1 Some values of accuracy obtained by cesium fountains 61
4.2 Uncertainty of a few other primary frequency standards 61

6.1 Noise processes involved in the frequency fluctuations of ultra-stable oscillators 83
6.2 Bandwidth of Allan, Hadamard and three-samples variances . . 112
6.3 Response to some noise spectral density laws 113
6.4 Sensitivity to a constant frequency drift 119
6.5 Cut-off sampling time 120
6.6 Typical values of the Allan standard deviation of the CHI-75 active hydrogen maser 124
6.7 Specification values of the Allan deviation for the 5071A primary frequency standard (high performance) 125

7.1 The uncertainties of the best realizations of the seven SI base units 129
7.2 Primary frequency standards involved in the accuracy of TAI . 133
7.3 Dates of the leap seconds since 1972 136
7.4 Countries maintaining a local approximation of UTC and/or an independent local time scale 138

1 Introduction

Frequency is the physical quantity that allows the most precise measurements. Time is a quantity dual of frequency: the unit of time is – and has always been – defined by a "periodical" phenomenon: its definition connects it to the duration of the period of the chosen phenomenon. It is then possible to measure the duration of any phenomenon, which then is expressed as the number of units it contains.

Conversely, the frequency of a periodic phenomenon is obtained by merely counting the number of periods occurring during the unit of time.

Giving such a definition to the unit of time poses a difficulty: a periodic phenomenon is one that repeats identically to itself at identical time intervals. Consequently, a time reference must be defined beforehand to decide whether a phenomenon is periodic, which means that a time reference should be known before defining a time reference!

In fact, an empirical and pragmatic approach has always been taken and this relies on the *assumption* that an observed phenomenon is periodic.

A unit of time is not sufficient to control the time parameter of a given activity. It must be complemented by a mean that allows a value of the duration of the time interval between any event and a conventional origin to be given. This time interval defines the *date* of the event, and the means of dating events is a *time scale.*

Since very early times, due to its direct practical implications in everyday life, the rotation of the Earth has been chosen to define the unit of time and the time scale. This choice has led to the time scale UT, with its two versions UT0 and UT1 now in use.

In this time scale, the basic period is the day, which is the period of the rotation of the Earth on its axis. The definition of the day and of the associated time scale has evolved with the improvement of the precision of the measurements and the requirements of the applications have become increasingly demanding:

1. In a first step, the local solar day (for a given location on the Earth) was merely the interval between two successive passages of the sun on the local meridian line (this is the so-called *true solar day*). This basic unit is divided into 24 hours, each hour having 60 minutes, etc.
   Noon (the passage of the sun on the local meridian line) was the origin of these time scales. As a consequence, there were as many different time

scales as municipalities located on different meridian lines and able to set a clock. Anyone traveling was consequently obliged to frequently reset his watch, unless he was traveling on a given meridian line.

2. Efforts have been made in two directions:
   a) Standardization of the time scale. All clocks in a given region (larger than a given municipality: a country, for instance) used the same time scale. For instance, in France, a law decreed in 1891 defined the legal time everywhere in France as being the solar time measured in Paris. Moreover, all clocks within a given region using a same time scale would differ by an integer number of hours from those in other zones using other time scales. The local time at the Royal Greenwich Observatory in England was chosen as standard at the 1884 International Meridian Conference [1], which stated:
      – Resolution IV: "*That the Conference proposes the adoption of a universal day for all purposes for which it may be found convenient, and which shall not interfere with the use of local or standard time where desirable.*"
        This resolution was adopted by the following vote: ayes, 23; abstaining, 2.
      – Resolution V: "*That this universal day is to be a mean solar day; is to begin for all the world at the moment of mean midnight of the initial meridian, coinciding with the beginning of the civil day and date of that meridian; and is to be counted from zero up to twenty-four hours.*"
        This resolution was adopted by the following vote: ayes, 15; nos, 2; abstaining, 7.

      The unit of time so defined is called a *true solar day*.
      In the United States and Canada, standard time zones were introduced on November 18, 1883; the French did not adopt the Greenwich meridian until 1911 (all the clocks in France had then to be adjusted by 9 min 21 s). By 1929 all major countries had adopted the time zones.
      Time scales in use to day are discussed in Sect. 7.1.
   b) Accuracy and stability of the time unit. It was quickly established that the unit of time "true solar day" was not a good reference: its length is not constant and this fact was easily shown using mechanical clocks, which were becoming increasingly stable. The length of a solar day varies throughout the year for different reasons:
      First, the Earth's orbit is an ellipse and the Earth moves faster when it is nearest the Sun and slower when it is farthest from the Sun.
      Second, the Earth's axis tilts, so that the Sun does not move exactly along the Earth's celestial equator but at a small varying angle above or below the celestial equator. Consequently, it moves faster or slower depending on whether it is moving toward or away from the equator.

The result of this complex apparent motion of the sun is that it crosses a given meridian with a delay or an advance of time that changes every day. The true solar day is shorter from February to March and from August to September and longer from May to June and from November to December. The difference between the longest and the shortest true solar day is about half an hour and was detected very early on.

The true solar day was replaced by the *mean solar day* as the basic unit of time. This new unit is defined as the mean value of the solar day calculated in a year. It equals the true solar day corrected by a value called the equation of time. This mean solar day was the basis of the second and of the UT1 time scales until 1956. Its implementation relies on astronomical observations.

In fact, the mean solar day is not constant either, for various reasons such as the tidal dissipation of energy, which slows down the rotation of the Earth, the atmospheric motions, the relative motion of the Earth core and of the Earth crust, etc.

The mean solar day was consequently abandoned in 1956, when the Comité International des Poids et Mesures decided that "*The second is the fraction 1/31 556 925.9747 of the tropical year for 1900 January 0 at 12 hours ephemeris time.*"

This new time scale (ephemeris time ET) is based on the rotation of the Earth around the Sun, which is much more stable than the rotation of the Earth about its axis. The origin of this time scale was chosen so that it coincided approximately with UT1 in the year 1900. Due to the slowing down of the rotation of the Earth, the difference between ET and UT1 was 56 seconds in 1988.

The most recent step in this effort to obtain a more efficient time scale is the use of atoms to define the unit of time, instead of astronomical objects.

The underlying assumption is that the energy levels of every atom of a given specie (for instance every atom of cesium, or every atom of hydrogen), if it is protected from perturbation, is the same everywhere in the universe and will never be modified, contrary to the motion of the Earth.

Since the energy levels of atoms define the frequency of the radiation emitted or absorbed when the atom undergoes a transition between these levels, this assumption implies that the frequency of the transitions of any atom of a given specie is the same everywhere in the universe and will never be modified.

Of course, many precautions must be taken when probing the true frequency of the chosen transition: any atom can be perturbed by its environment and by the probing process, with the result that the energy levels are modified. These modifications must be minimized and taken into account.

One of these transitions can consequently be chosen as the periodic phenomenon that defines the elementary time interval: a transition between two energy levels of the ^{133}Cs atom has defined the second since 1967 (13th Conférence générale des Poids et Mesures [35]):

"*The second is the duration of 9 192 631 770 periods of the radiation corresponding to the transition between the two hyperfine levels of the ground state of the cesium 133 atom.*"

This decision was prepared by the progress made in the atomic frequency sources.

The fractional accuracy and the fractional stability attained by the best realization of the time SI unit are now (2006) better than one part in 10^{15}.

Excellent approximations of the realization of this new definition of the SI time unit can be made by any institute of metrology in the world.

Moreover, reference oscillators with very high qualities can be widely spread among the users and thus allow a large variety of applications that benefit from their extremely good properties. Telecommunications, global positioning, very long base interferometry, and physical research are some of these applications. Some of them will be described in this book.

Handling quantities with such accuracy and stability is specific to the time-frequency domain and special care must be taken in order to not corrupt these qualities. The aim of this book is to describe and study the procedures that allow one to obtain the best result from this unique and extraordinary tool that is time and frequency.

In Part I, the conception of the frequency standards (oscillators that deliver a frequency that can be used as a reference) is described. The best one is, of course, the atomic frequency standard, whose output frequency is controlled by an atomic transition. Nevertheless, the principle of an atomic frequency standard is basically the same that of any other frequency source.

A high quality oscillator consists of two parts that interact with each other as follows:

- In one part, a physical phenomenon is produced, which involves the reference frequency. For instance, in the case of an atomic oscillator, the atoms of interest are produced in the desired energy level and the transition of interest is excited.
- In the other part, the oscillation to be controlled by the physical phenomenon is produced by an electronic (or optical) device and compared to the frequency of the physical phenomenon. This comparison implies an interaction between the oscillation and the phenomenon. This results in an error signal, which allows one to maintain the frequency of the oscillation at a value related to the value of the reference frequency.

These two parts are detailed in Chap. 2 and Chap. 3.

The frequency of the output of the overall device is a normalized one (for instance 10 MHz, 1 Hz, etc.) in order to be easily handled by the end-users.

A clock is a device that generates periodic, accurately spaced signals used for timing applications and counts the number of periods, for instance the number of seconds, minutes, etc., that have elapsed since a conventional origin. It is consequently able to give the date of an event.

A clock consists of at least three parts:

- an oscillator;
- a device that counts the oscillations and converts them to units of time interval (such as seconds, minutes, hours, and days);
- a means of defining the origin; and
- a means of displaying or recording the results.

Consequently, oscillators underlie all time-frequency activities.

The most representative atomic oscillators are described in Part I of the book (see Sect. 2.3). They are:

- the active hydrogen maser,
- the frequency standard using an atomic beam of cesium, and
- the frequency standard using an atomic fountain of cesium.

A precise characterization of the devices is necessary in order to determine the types of noise that limit the accuracy of the measurements to be performed, as well as the performances of applications relying on its use.

The metrological properties of interest that must be known in most applications are the stability, the accuracy and the reproducibility.

The accuracy of an oscillator is the degree of conformity of the frequency of the signal it delivers to its nominal value.

The reproducibility is correlated to the accuracy: it indicates how the output frequency of a device of a given kind (e.g. the cesium atomic frequency standard 5071A) differs from that of another device of the same type.

The evaluation of the accuracy, which is often called uncertainty, is based on a careful and precise review of the physical effects that can modify the frequency of the reference phenomenon (i.e. the atomic reference transition, often called the clock transition) and of their uncertainty (budget error).

This kind of review is described in a precise example in Chap. 4.

While accuracy indicates whether the value of the frequency is right or wrong, and how far wrong it can be, stability indicates whether or not this value stays the same over a given time interval.

The value of the frequency of an oscillator varies with time due to:

1. ergodic and second-order stationary random fluctuations, and
2. non-stationary random or deterministic variations of the frequency.

Two different approaches are used to characterize the stability:

1. Direct measurements of the spectrum of the signal delivered by the oscillator in frequency domain. Of course, the local oscillator must be at least as good as the oscillator under test. Such measurements are well suited to characterize the effect of the stationary random fluctuations far from the carrier.

2. Calculation in the time domain of the variance of the results of the measurement of the frequency of the oscillator. Various numeric filters are applied to the sequence of results of these frequency measurements, which lead to different variances: Allan, Hadamard and three-samples variances are the most used.
   Such measurements are well suited to characterize the effect of the stationary random fluctuations close to the carrier, and one of them (the three-samples variance) gives a result free from any stationary constant drift.
   These algorithms are described in Chap. 6.

Part III of the book describes three important applications of precise frequency standard, where the best qualities are required:

- Frequency and time metrology.
  The first objective of metrology is to distribute the SI unit to the user in order to insure coherent measurements. In the case of time and frequency, the high performance frequency and standards disseminated all over the world are used to define the different time scales in use.
  A time scale is a kind of virtual clock defined by a set of real clocks, whose goal is to keep time and provide it for the end-users.
  It uses a set of frequency sources and an algorithm to construct the time scale from the data delivered by the clocks participating in it and an agreement to define the length of the second (which is the standard unit of time interval). Seconds are then counted to measure longer time intervals, such as minutes, hours, days, etc.
  These time scales are described herein. Moreover, a method allowing a very accurate comparison of the clocks taking part in the time scale construction, as well as the comparison of any user clock with the time scale, are discussed.
- Tracking and localization.
  One of the most popular applications of precise time and frequency is tracking and navigation, particularly by the wide use of global systems such as the global positioning system (GPS) and the future GALILEO.
  GPS uses a constellation of satellites that orbit the Earth. By processing signals received from the satellites, a commercial GPS receiver can determine its own position within an uncertainty of a few meters.
  GPS is also a dominant system for the distribution of time and frequency. There are several types of time and frequency measurements that involve GPS, including one-way, common-view, and carrier-phase measurements.
  The space segment, control segment and user segment of GPS are described and the processing of the data transmitted by the satellites is studied and discussed.
- Very long base interferometry (VLBI).
  This measurement technique relies on the observation of the time difference between the arrival at two or more antennas of a radio wavefront

emitted by a distant radio source. The distance between the antennas is the base of the interferometer.

The more distant the antennas, the more precise the measurements. VLBI is an interferometry technique where the base is of the order of a few thousands kilometers. Consequently, the signals received by the antennas are not correlated in real time, but recorded and sent to a common correlator where they are played together.

This is only possible if a very stable time reference is used at each antenna. First, the local oscillator that down converts the frequency of the received signal to the IF frequency must have a very low phase noise, so that the fringes are not blurred. Second, precise timing information must be recorded with each signal in order to synchronize them in the correlator and deduce the time delays of the wavefront at each antenna.

Part I

Very High Performances Oscillators

Since the second half of the 20th century frequency standards and clocks with increasingly better qualities have been designed and operated:

- In 1948, the National Bureau of Standards (NBS, now the National Institute of Standards and Technology NIST) built the first atomic (molecular) clock (see [86, 87]). It used an absorption line of molecules of ammonia and the fractional accuracy of this very first device was one part in 10^7. This frequency source was never used for timekeeping because its properties were not better than those of quartz clocks.
- The first accurate atomic clock was based on a transition of the cesium atom and was built in 1955 at the National Physical Laboratory [46]. The first version of this device was accurate to one part in 10^{10}.
- The first commercially available cesium clock, the Atomichron, went on sale in October 1956 [77]. From 1956 to 1960 about fifty of these devices (NASF/NC-1001 Atomichron) were made and sold [48].

These devices launched the world into a new concept of timekeeping, leading to a new definition of the SI time unit and to increasingly accurate and stable devices, which allowed new applications.

The principle of these high performance oscillators – frequency standards – will be studied in the following chapter, and some of the best ones will be described. This principle can be applied to optical oscillators, i.e. oscillators whose output frequency is in the optical domain. These oscillators are not described in this book.

Processing of the atomic line used as the frequency reference (clock transition) can damage the quality of the device if it is not carefully designed. A careful analysis of this part of any frequency standard is therefore given in the following chapter.

2 The Two Parts of an Oscillator

Any electronic or optical oscillator is composed of two parts whose quality must be optimized to obtain the best performances (see Fig. 2.1). The two parts are as follows:

1. A package where a periodic physical phenomenon is produced. The frequency of this phenomenon is used by the other part of the device to control the frequency of the output signal of the oscillator. This is the reference frequency of the device.
2. An electronic and/or optical package. This part controls the periodical phenomenon and its environment and links the frequency of the output signal of the oscillator to that of the physical phenomenon.

The quality of the oscillator consequently depends on the following two types of factors:

1. The factors related to the physical phenomena involved in the first part of the device. The stability, accuracy, etc. of the device cannot be better than that of the physical phenomenon.
2. The factors related to the electronic and/or optical part whose task is to control and interrogate the physical phenomenon and then to process the data resulting from this interrogation. Noise and errors that are introduced by these processes lower the quality of the output signal and must consequently be minimized.

A few examples of oscillators, using this scheme, are given below. They differ by the nature of the physical phenomenon involved.

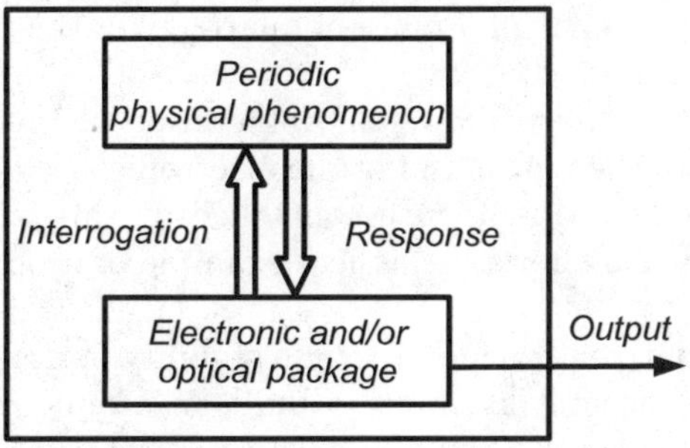

Fig. 2.1. Block-diagram of an oscillator

2.1 Electrical Resonances

The physical phenomenon used in simple electronic oscillators is the electrical resonance of a selective circuit, such as the L-C parallel circuit of the classical Robinson oscillator shown in Fig. 2.2. This book is devoted to high performance oscillators and does not discuss this family of oscillators.

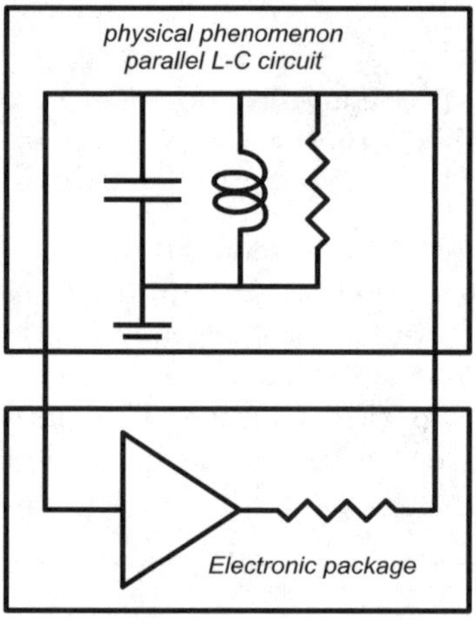

Fig. 2.2. Robinson oscillator

2.2 Mechanical Resonances

A wide family of oscillators uses mechanical vibration of piezo-electric material.

When a piezo-electric element is stressed electrically by a voltage, its dimensions change. When it is stressed mechanically by a force, it generates an electric charge. If the electrodes are not short-circuited, a voltage associated with the charge appears.

The vibration frequency of a plate of quartz gives the reference frequency of the oscillator.

This piezo-electric effect allows the interaction of the physical phenomena (the vibration of the plate) and of the electronic circuit, which drives the vibration of the plate and is in turn controlled by this vibration via a feedback circuit. Figure 2.3 shows a classical example of a quartz crystal Pierce oscillator.

Quartz oscillators can provide very high stability performance. One should bear in mind that the first attempt to built an atomic frequency standard did not give a better result than the existing quartz crystal oscillators (see p. 11).

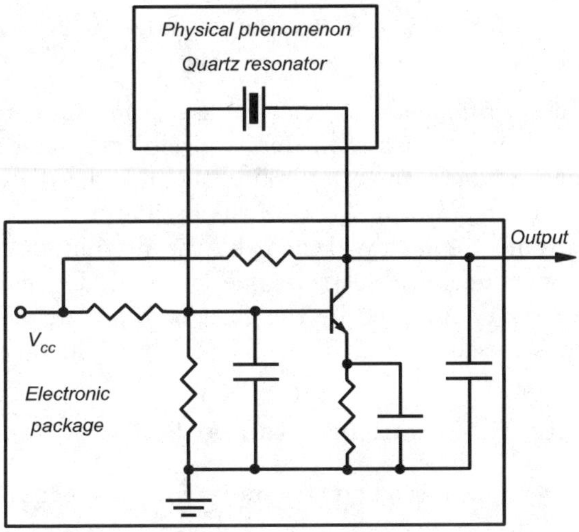

Fig. 2.3. Pierce oscillator

2.3 Atomic Frequency Standards

In this case, the physical phenomenon is an atomic resonance, e.g. a transition between two levels of a given atomic specie.

Every atom possesses an infinite number of discrete states depending on the configuration of its electrons. Each state is characterized by a precise energy, the state with the lowest energy being the ground state. Any transition between two states modifies the energy of the atom and must consequently be accompanied by an energy exchange with its environment. This exchange often occurs by emission or absorption of an electromagnetic radiation whose frequency f is defined by the energy difference δE between the two states involved in the transition:

$$\delta E = h \times f \;, \tag{2.1}$$

where h is the Planck constant whose value is (see [94])

$$h = 6.626\,0693 \times 10^{-34}\,\mathrm{J \cdot s} \;. \tag{2.2}$$

Moreover, each state may possess fine and hyperfine structures. A fine structure arises because the electron has an intrinsic angular momentum (spin) that interacts with the magnetic field produced, for example, as the electron "orbits" the nucleus. A hyperfine structure is produced by the interaction of the electronic spin and that of the nucleus. It depends on the spin of the electrons and of the nucleus of the atom.

Each state of an atom is characterized by a set of values, the quantum numbers. The atomic transition that is chosen for a given atomic frequency standard is its "clock transition".

A distinction is often made between active and passive atomic frequency standards as follows:

- In active atomic frequency standards, the atoms undergo a transition toward a lower energy state and emit an electromagnetic signal at the frequency of the clock transition. This signal is used by the electronic package to build the output signal of the oscillator.
- In passive atomic frequency standards, the atomic cloud is too dilute and does not produce any electromagnetic signal. The clock transition is probed by a signal generated by the electronic package. The response to this excitation is used to control the frequency of the output signal.

2.3.1 Active Atomic Frequency Standards

The active hydrogen maser is a good example of an active frequency standard.

Its clock transition connects the two hyperfine levels of its ground state. These levels are designed by the quantum numbers $F = 1$, $m_{\mathrm{F}} = 0$ (the state with the higher energy) and $F = 0$, $m_{\mathrm{F}} = 0$ (the state of the lower energy). The energy difference between these two states corresponds to the frequency $f_{\mathrm{H}} = 1\,420\,405\,752\,\mathrm{Hz}$.

A very dilute cloud of hydrogen atoms is confined for about one second in a resonant cavity tuned to the frequency f_{H} of the clock transition.

The atoms are sorted before their introduction into the cavity: only those in the upper state $F = 1$, $m_{\mathrm{F}} = 0$ of the hyperfine structure enter the cavity. They coherently amplify the noise component at the frequency f_{H} of the hyperfine transition. If the atomic density is high enough, an oscillation at that frequency is produced in the cavity and can be detected (this is the maser effect[1]).

The electronic package controls *the phase* of the local 5 MHz or 10 MHz oscillator in such a way that it is linked to the phase of the atomic signal.

Figure 2.4 shows a schematic block diagram of the hydrogen maser.

2.3.2 Passive Atomic Frequency Standard

The cesium beam frequency standard is a good example of passive atomic frequency standard.

In this case, the atoms are not confined. They form an atomic beam and are consequently too dilute to produce the maser effect. Consequently, there is

[1] Maser: microwave amplification by stimulated emission of radiation. The principle of stimulated emission was proposed by Albert Einstein in 1917. When atoms have been put into an excited energy state, they can amplify incoming radiation at the frequency of the transition between the present state and the fundamental state by emitting a photon coherent with the incoming radiation (i.e. corresponding to an electromagnetic wave with same phase and same polarization as the incoming radiation).

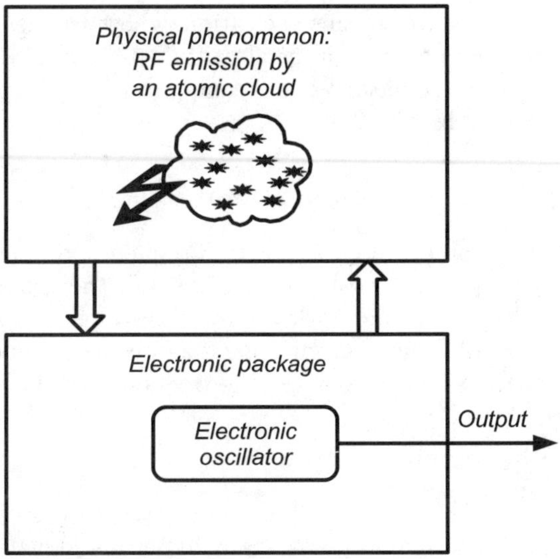

Fig. 2.4. Block diagram of an active hydrogen maser

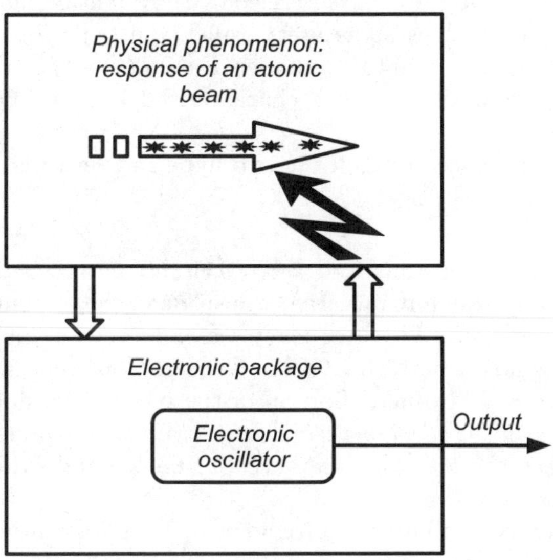

Fig. 2.5. The two parts of a cesium beam frequency standard

no emission of a radiation that can be used to lock the phase of the electronic oscillator.

An excitation, produced by frequency synthesis from the electronic oscillator, probes the atomic transition that connects the two hyperfine levels $F = 4$, $m_{\mathrm{F}} = 0$ and $F = 3$, $m_{\mathrm{F}} = 0$ of the ground state of the cesium atom (isotope 133). This excitation is the interrogation of the atoms.

The response of the atoms to this excitation, whose frequency is related by a known number to the frequency of the electronic oscillator, allows correction of the frequency of this oscillator via a frequency control loop.

This effect is shown in Fig. 2.5.

2.3.3 Optically Pumped Cesium Beam

The physical phenomenon that provides the frequency reference is the resonant transition between the two hyperfine levels of the ground state of the cesium atom. The electronic and optical part of the device probes the atomic transition in a beam of cesium atoms that has been suitably prepared and uses the answer of the atoms to correct the frequency of the output.

The Physical Phenomenon

The ground state of the cesium atom has no fine structure because the orbital magnetic moment cancels, but possesses a hyperfine structure due to the interaction of its nucleus spin (characterized by the quantum number $I = 7/2$) and its valence electron spin (characterized by the quantum number $S = 1/2$). If the electron and nucleus spins are parallel, the hyperfine structure level is characterized by the quantum number $F = 7/2 + 1/2 = 4$. If they are anti-parallel, the hyperfine level is characterized by the quantum number $F = 7/2 - 1/2 = 3$.

The energy difference between the two hyperfine levels is

$$\Delta E = h \times f_{\mathrm{Cs}} \,, \tag{2.3}$$

where h is the Planck constant, and $f_{\mathrm{Cs}} = 9\,192\,631\,770$ Hz is the frequency of the radiation involved in any radiative transition of the cesium atom between the two hyperfine levels. This frequency was adopted in 1967 to define the SI time unit (second). The 13th CGPM (General Conference on Weights and Measures) gave the following definition of the second [35, 66][2]: *"The second is the duration of 9 192 631 770 periods of the radiation corresponding to the transition between the two hyperfine levels of the ground state of the cesium 133 atom".*

Consequently, the value of the frequency f_{Cs} is known without any uncertainty[3].

[2] The reference text is in French and reads: "La seconde est la durée de 9 192 631 770 périodes de la radiation correspondant à la transition entre les deux niveaux hyperfins de l'état fondamental de l'atome de césium 133".

[3] Of course, this frequency is the frequency of *unperturbed* cesium atoms. For instance, at its 1997 meeting, the CIPM (Comité International des Poids et Mesures – International Committee for Weights and Measures) confirmed that *"This definition refers to a cesium atom at rest at a temperature of 0 K"* in order to make it clear that the cesium atom must not be perturbed by black-body radiation and that the hyperfine frequency must be free of any Doppler effect.

The basic idea in order to lock the frequency f_o of an electronic oscillator (5 or 10 MHz, for instance) to the frequency f_{Cs} of the transition is as follows.

1. Produce a beam of cesium atoms.
2. Produce an unbalance of the populations of the atoms in the two hyperfine levels $F = 3$ and $F = 4$, which otherwise are almost the same at thermal equilibrium. These two steps are the preparation of the beam.
3. Generate a signal, whose frequency f_1 is close to the frequency f_{Cs}, by frequency synthesis from the output of the electronic oscillator.
   The ratio f_1/f_o is exactly known and results from classical methods of frequency addition, subtraction, multiplication and division.
4. Excite the beam of atoms by this signal of frequency f_1. The result of this excitation, if the frequency f_1 is close to the frequency f_{Cs}, is to balance the populations of the $F = 3$ and $F = 4$ atoms.
   This is the interrogation of the beam.
5. Measure the residual population unbalance after the interrogation, which indicates how far f_1 is from f_{Cs}. This is the response of the beam.
6. The response of the beam is then used to correct the frequency f_o, so that it is as close as possible to its nominal value according to the definition of the SI second.
   This is the servo control of the output frequency.

These steps can be found in every passive atomic frequency standard.

They are summarized in Fig. 2.6 and details are given below.

The atomic beam is in a vacuum chamber as good as possible, and the region where the atoms are excited by the microwave radiation is protected from spurious magnetic field by magnetic shields.

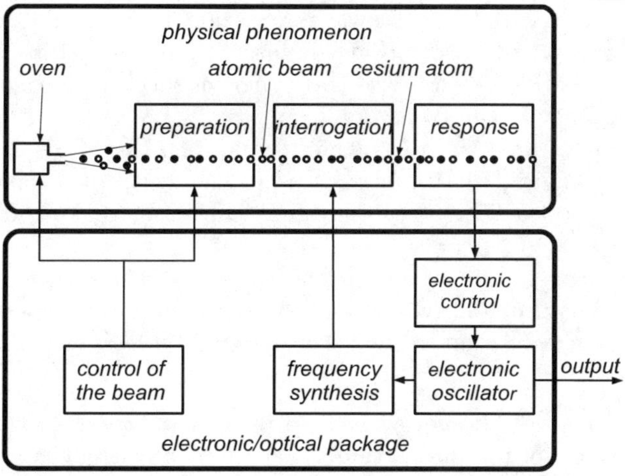

Fig. 2.6. Block diagram of a cesium beam frequency standard

Preparation of the Atoms

Two set-up families have been designed to prepare the atomic beam and produce the population unbalance.

1. Magnetic deflection of the atoms.
   Historically, this was the first scheme. The interaction between an atom and an inhomogeneous magnetic field is different according to the hyperfine state of the atom [51].
   A magnet producing an inhomogeneous magnetic field (Stern–Gerlach magnet) spatially separates atoms in the various mF states, and atoms in one of the ground-state levels ($F = 3$, $m_{\mathrm{F}} = 0$ or $F = 4$, $m_{\mathrm{F}} = 0$) are transmitted through the interrogation region. The other atoms are lost.
   The response of the atoms to their interrogation is detected by an identical Stern–Gerlach magnet arranged so that atoms are directed to a detector only if they have been stimulated by the microwave field to the other ground, $m_{\mathrm{F}} = 0$, level.
   This scheme is still used in the commercially available cesium beam frequency standards.
2. Optical pumping of the atoms.
   The state of the atoms is manipulated by optical radiations. All the atoms of the beam can in principle be "pumped" in one of the two hyperfine states. The response of the atoms is also detected by their interaction with an optical beam.
   The present primary cesium beam frequency standards use this optical pumping scheme.

Excitation of the Transition

The atoms interact with the microwave excitation in a resonant cavity (tuned to the frequency f_{Cs}) in order that the interaction strength be sufficient to produce the transition between the two hyperfine levels if the excitation is at the exact frequency of the transition.

In the case of the cesium frequency standards, the cavity has a special shape (Ramsay cavity [106]): every atom interacts with the excitation in two successive regions, each having a length l. These two regions are separated by a distance L.

What happens to the atoms in the cavity is well known (see, for instance [123]). The probability that a given atom in a given hyperfine $m_{\mathrm{F}} = 0$ level undergoes a transition to the other hyperfine $m_{\mathrm{F}} = 0$ level in the cavity (transition probability) depends on the following.

- The amplitude of the microwave field in the cavity.
- The duration t of the interaction in each part of the cavity.
- The delay T between the two interactions.

- The difference $\delta f = f_1 - f_{\mathrm{Cs}}$ between the microwave excitation frequency f_1 and the resonance frequency f_{Cs}.

This transition probability is given by

$$P(b, \delta\omega, t, T) = \frac{4b^2}{\Omega^2}\sin^2\left(\frac{\Omega t}{2}\right) \times \left[\cos\left(\frac{\Omega t}{2}\right)\cos\left(\frac{\delta\omega T + \Phi}{2}\right)\right]^2 - \frac{4b^2}{\Omega^2}\sin^2\left(\frac{\Omega t}{2}\right) \times \left[\frac{\delta\omega}{\Omega}\sin\left(\frac{\Omega t}{2}\right)\sin\left(\frac{\delta\omega T + \Phi}{2}\right)\right]^2 . \tag{2.4}$$

In this expression

- $\delta\omega = 2\pi\delta f$ characterizes the mistuning of the excitation and the parameter,
- b (the so-called Rabi frequency) characterizes the amplitude of the microwave excitation

$$b = \frac{\mu_{\mathrm{B}}}{\hbar}B = \frac{e}{2m_{\mathrm{e}}}B , \tag{2.5}$$

where B is the microwave magnetic field amplitude in the cavity, μ_{B} is the Bohr magneton:

$$\mu_{\mathrm{B}} = e\hbar/2m_{\mathrm{e}} = 9.27400949(80) \times 10^{-24}\,\mathrm{J}\cdot\mathrm{T}^{-1} ,$$

e is the elementary charge, $\hbar$ is the reduced Planck constant:

$$\hbar \equiv \frac{h}{2\pi} = 1.05457168(18) \times 10^{-34}\,\mathrm{J}\cdot\mathrm{s} ,$$

m_{e} is the electron rest mass,
- $\Omega = \sqrt{\delta\omega^2 + b^2}$, and
- Φ is the phase difference of the microwave between the two parts of the cavity.

Figure 2.7 shows the calculated transition probability of an atom whose speed is such that the interaction time is $t = 50\,\mu\mathrm{s}$ (Ramsey fringes).

This calculation has been made for the following values: $L/l = 10$, $b = 10^4\,\mathrm{Hz}$, $\Phi = 0$, $-10\,\mathrm{kHz} \leq f_1 - f_0 \leq 10\,\mathrm{kHz}$.

This shows that the transition probability is maximal for $\delta\omega = 0$ (the central lobe of the curve) and that many side lobes exist, with smallest values of the transition probability. The maximal value of the central lobe (for $\Phi = 0$) is

$$p_{\max} = \sin^2(bt) \tag{2.6}$$

and its width is

$$\delta f = \frac{1}{2T} . \tag{2.7}$$

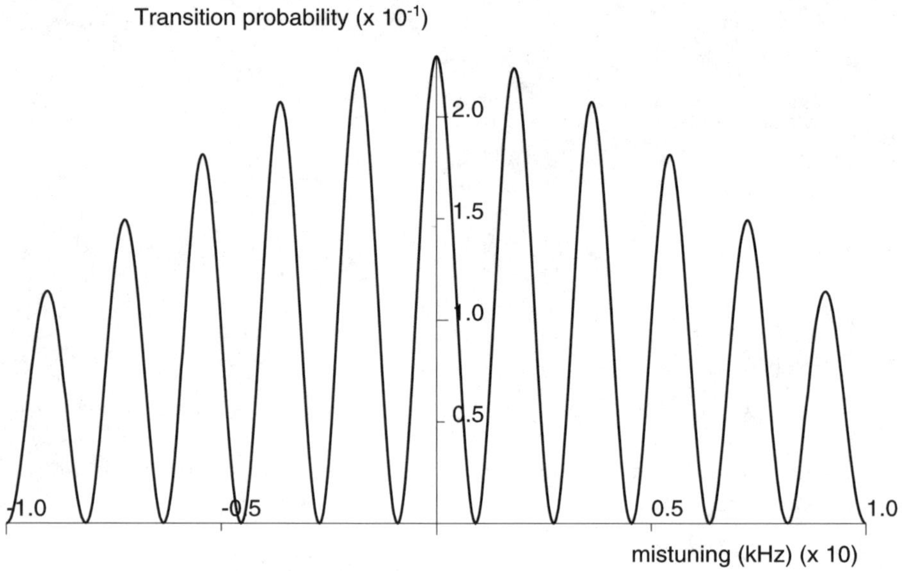

Fig. 2.7. Transition probability of an atom across a Ramsay cavity

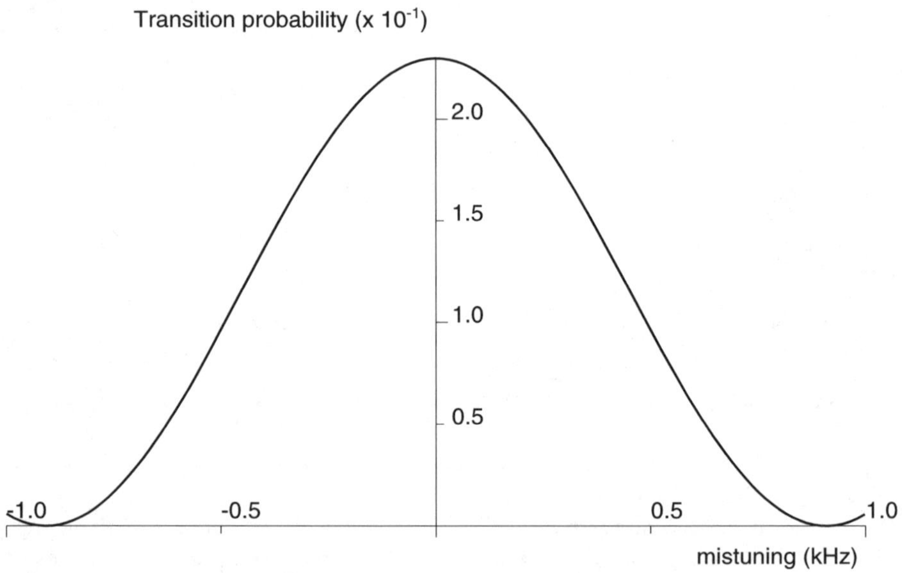

Fig. 2.8. Central part of the pattern of Fig. 2.7

This central lobe is used to lock the frequency of the local oscillator to the frequency of the transition, since its maximal value occurs for the exact value of the transition frequency. Figure 2.8 shows this central lobe: its full width is 1.0 kHz, which corresponds to $T = 500\,\mu s$.

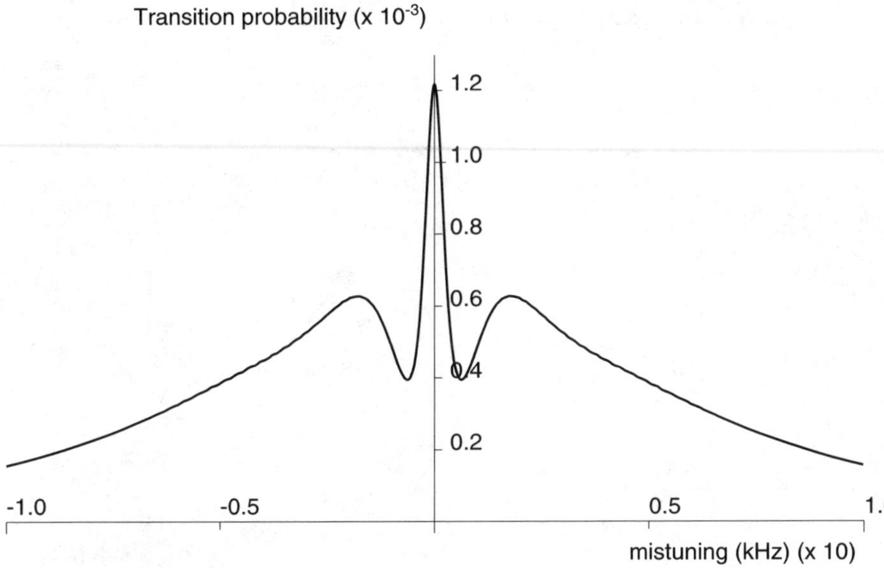

Fig. 2.9. Mean transition probability of a beam with a velocity distribution

In the case where $\delta\omega \ll b$ (the excitation frequency f_1 is very close to the resonance frequency f_{Cs}), (2.4) becomes

$$\begin{aligned} P(b, \delta\omega, t, T) &= 4\sin^2\left(\frac{bt}{2}\right)\cos^2\left(\frac{bt}{2}\right)\cos^2\left(\frac{\delta\omega T + \Phi}{2}\right) \\ &= \sin^2(bt)\cos^2\left(\frac{\delta\omega T + \Phi}{2}\right) . \end{aligned} \tag{2.8}$$

In fact, in a thermal beam, there is a distribution of the atomic velocities and only the central lobe remains visible, the other ones being blurred by the velocity dispersion.

Figure 2.9 shows the calculated mean transition probability of the atoms of a thermal beam with a Maxwell–Boltzmann velocity distribution. The most probable velocity of the beam corresponds to an interaction time of $t = 50\,\mu s$ and $T = 10 \times t$. Figure 2.10 shows the central part of the pattern in Fig. 2.9. Its full width is $\delta f = 630\,\text{Hz}$, which corresponds to an equivalent interaction time t of 79 μs.

Detection of the Atomic Transition

In order to lock the local oscillator frequency to the frequency of the atomic transition (multiplied by a known factor), it is necessary to measure how well the atoms have undergone the transition in the interaction region, i.e. to measure the transition probability.

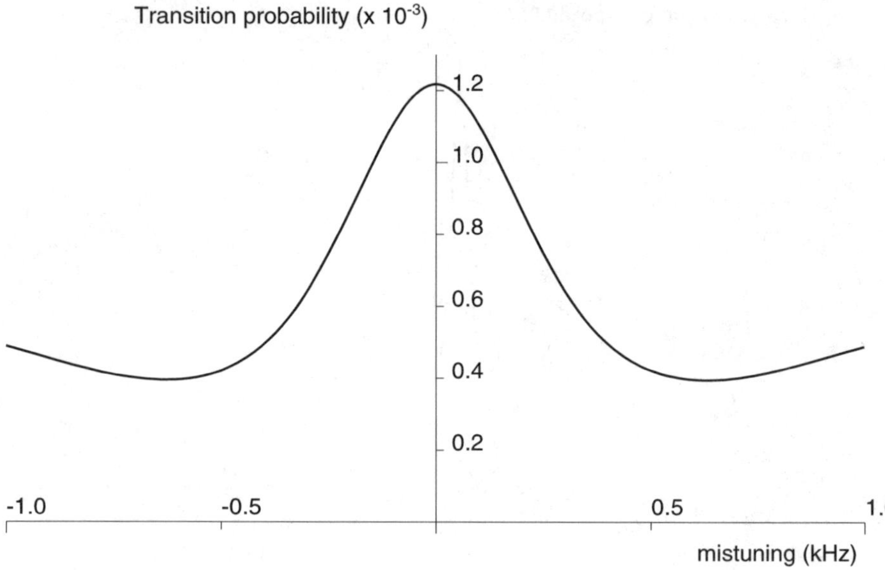

Fig. 2.10. Central part of the curve of Fig. 2.9

As for the preparation of the beam (see p. 20), two methods are used to detect the atomic transition.

1. Magnetic deflection of the atoms. This set-up was used in the cases where the beam preparation itself used magnetic deflection. Only the atoms in a given hyperfine state are directed toward a detector. The response of this detector consequently gives the number of atoms having undergone the transition.
2. Optical detection of the atoms. Only the atoms in a given hyperfine state interact with the laser beam; it is consequently possible to know the number of atoms having undergone the transition via the fluorescence signal produced by the atomic beam excited by this laser beam.

The Electronic and Optical Package

Besides controlling the experiment (vacuum system, oven producing the cesium beam, control of environmental parameters such as temperature and magnetic field), the main task of this part of the device is to control the frequency of an electronic oscillator (local oscillator, LO) in such a way that

- its frequency remains locked (but not equal[4]) as tightly as possible to the reference frequency (i.e. the frequency of the transition) in order to ensure accuracy and long-term stability;

[4] The frequency of the transition is imposed by the clock transition chosen for a given device, when the output frequency must be a standard frequency such as 10 MHz. Both are connected by a frequency synthesis as shown in the following.

– its frequency noise remains as low as possible to ensure as good a short-term and medium-term stability as possible.

Generation of the frequency $f_1 \approx f_{Cs}$

The signal at frequency f_1 is applied to the Ramsay cavity in order to probe the atomic transition. The atomic response to this excitation provides the error signal that allows locking of the local oscillator (frequency f_o) to the atomic frequency. The noise properties of the excitation signal must consequently be optimized.

Figure 2.11 shows the simplified block diagram of an electronic device that synthesizes this frequency f_1, close to the reference frequency f_{Cs} [90].

It is generated by frequency multiplication (overall multiplication factor: 184) and synthesis from the 5 MHz local oscillator. This local oscillator is a voltage controlled quartz crystal oscillator as shown in Fig. 2.11.

Another simplified example is shown in Fig. 2.12 (see [112]).

Both are simplified block diagrams that do not show the efforts made to reduce the phase noise of the excitation signal.

The main improvement to these general block diagrams in order to minimize the phase noise is the use of oscillators to produce some intermediate frequencies of the chain: the phase (and frequency) noise of a good 20 MHz

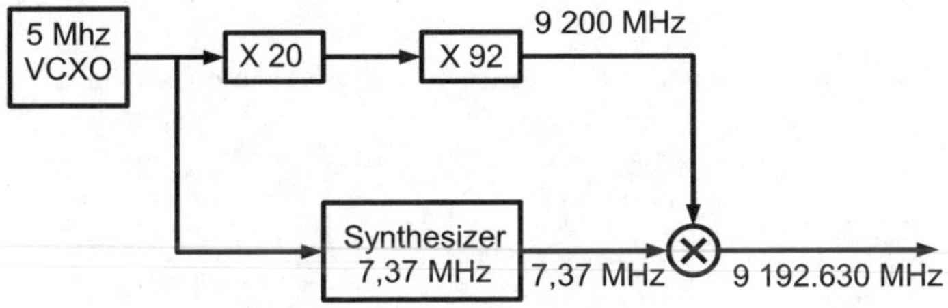

Fig. 2.11. Generation of the frequency f_1

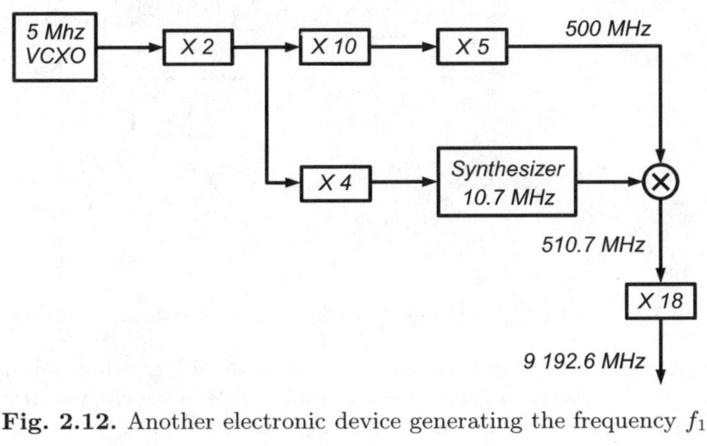

Fig. 2.12. Another electronic device generating the frequency f_1

oscillator (for instance) is better than the noise of the frequency-multiplied 5 MHz local oscillator.

These intermediate oscillators are phase locked to the local oscillator. In the case of [90], the final frequency itself (9192 MHz) is generated by a DRO[5] This oscillator is phase locked to the signal of frequency f_1 generated by frequency multiplication and syntheses from the local oscillator. This is shown in Fig. 2.13.

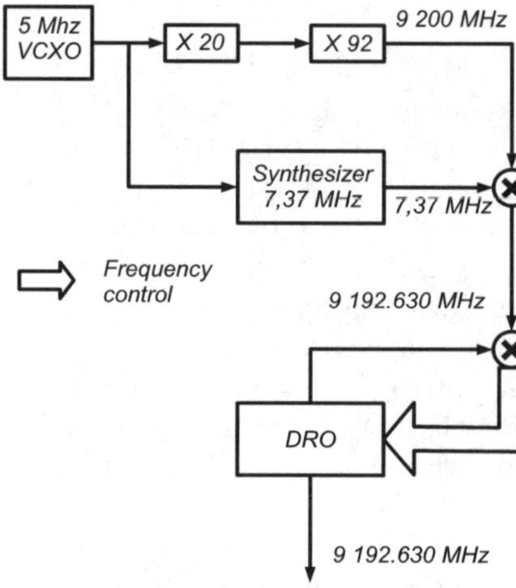

Fig. 2.13. The excitation of the Ramsay cavity as the output of a DRO [90]

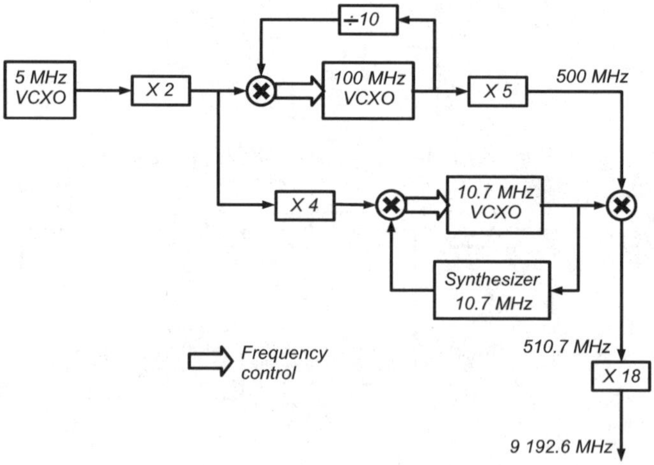

Fig. 2.14. Some frequencies generated by intermediate oscillators [112]

[5] DRO: dielectric resonator oscillator. This is an oscillator whose physical phenomenon is the electromagnetic resonance of a dielectric cavity. It provides low phase (and frequency) noise and high power output.

In the case of [112], the frequencies 100 MHz and 10.7 MHz are the intermediate frequencies generated by oscillators as shown in Fig. 2.14.

- The 100 MHz oscillator is phase locked to the 10 MHz output of the ×2 multiplier. To this end, the output of this 100 MHz oscillator is divided by 10.
- The 10.7 MHz oscillator is phase locked to the output of the frequency synthesizer driven by the output of the ×4 multiplier.

Locking the Local Oscillator Frequency to the Transition Frequency

The result of the excitation of the atoms in the Ramsay cavity is used to control the frequency of the local oscillator in such a way that its frequency remains as close as possible to its nominal value (5 MHz, for instance). For this purpose, an error signal is generated from the detection of the atomic transition in various ways.

1. In all cases, the frequency of the excitation is modulated in order to produce a periodic response that can be detected with a signal to noise ratio much better than for a static excitation (synchronous detection).
2. The frequency of the modulation is low enough so that the calculated static atomic response can be applied.
3. Some devices use a sinusoidal modulation of the frequency and synchronous detection of the third harmonic, which provides an error signal that cancels when the carrier frequency equals the frequency of the transition and whose sign is positive or negative according to what side of the atomic resonance curve the carrier frequency is.
4. Some devices use a square wave modulation. In this case, static approximation cannot be applied to the atomic response. The transient part of the response, which occurs at every half period of the modulation, is not taken into account and only the constant (or slowly varying) part of the response is used to generate the error signal. A digital processing is well suited for this purpose.

The error signal may be used

- either to permanently correct the frequency of the local oscillator. In this case, the feedback loop is closed and the device delivers that frequency controlled by the transition frequency;
- or to record the fluctuations of this local oscillator relative to its nominal frequency. In this case, the feedback loop is open.

2.3.4 Active Hydrogen Maser

This device uses the maser effect (see p. 16) in a very dilute gas of atomic hydrogen [52, 71].

In this gas, the atoms are in the upper hyperfine state of their ground level: they emit a coherent radiation at the frequency of the hyperfine transition

(1 420 405 752 Hz). The local oscillator of the device is phase-locked to this radiation.

The Physical Phenomenon

The physical phenomenon is the maser effect, which produces a weak electromagnetic radiation whose frequency is defined by the energy difference between the two hyperfine levels of the ground state of the hydrogen atom.

As in the case of the cesium atom, the ground state of the hydrogen atom has no fine structure because the orbital magnetic moment cancels, but possesses a hyperfine structure due to the interaction of its nucleus spin (characterized by the quantum number $I = 1/2$) and its valence electron spin (characterized by the quantum number $S = 1/2$).

If the electron and nucleus spins are parallel, the hyperfine structure level is characterized by the quantum number $F = 1/2 + 1/2 = 1$. If they are anti-parallel, the hyperfine level is characterized by the quantum number $F = 1/2 - 1/2 = 0$.

In order to produce the maser effect, a population difference must be permanently produced in a collection of atomic hydrogen: the number of atoms in the upper energy hyperfine level ($F = 1$) must be greater than the number of atoms in the lower hyperfine level ($F = 0$).

Moreover, because the microwave power produced by the stimulated emission is very weak, the maser effect must occur in a resonant cavity tuned to the hyperfine frequency.

The overall scheme of the part of the device producing the maser effect is shown in Fig. 2.15. It shows the following.

1. The atomic hydrogen source. Atoms are produced by an RF discharge in a cell containing molecular hydrogen at a low pressure (1 Pa).
2. The magnetic selection. An inhomogeneous magnetic field produced by a hexapolar magnet along the beam axis allows selection of the atoms according to their hyperfine state. The atoms in the upper ($F = 1$) hyperfine state are recalled toward the axis of the device, while the atoms in the lower ($F = 0$) hyperfine state are repelled out of the axis; only the atoms in the upper level are focused on the entrance of a teflon[6]-coated bulb.
3. The region where the dilute cloud of atoms with a permanent population difference stands. The teflon-coated quartz bulb is located inside the cavity tuned to the hyperfine transition frequency. The atoms in the hyperfine state $F = 1$ that have entered this bulb stay in it for about

[6] Teflon is the brand name of a polytetrafluoroethylene (PTFE), a polymer of fluorinated ethylene. It is very non-reactive, and only very weakly perturbs the hydrogen atoms undergoing a collision on the wall of the coated bulb. DuPont patented it in 1941 and registered the Teflon trademark in 1944.

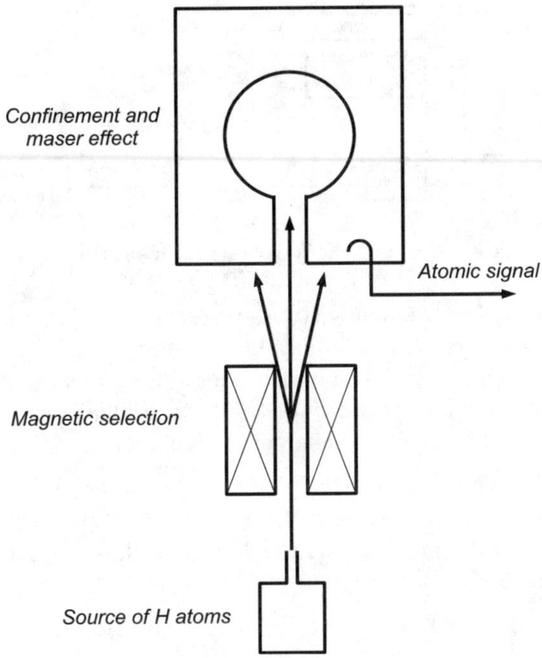

Fig. 2.15. Production of the maser effect

one second and emit a photon at the transition frequency by stimulated emission.
The atomic radiation is coupled to the output by a loop or an antenna in the cavity. The power of this output signal is of the order of 10^{-13} W. This is enough to insure a good phase locking of the local oscillator (5 MHz, for instance) to its phase.

The atomic beam is in a vacuum chamber that is as good as possible, and the region where the atoms emit the maser radiation is protected from spurious magnetic field by magnetic shields.

The width of the emitted line is limited by the time of the interaction of the atoms and of the radiation itself, which is of the order of one second.

The Electronic Package

The functions of this part of the device are

1. to control the physical phenomenon production: the atomic hydrogen source, the vacuum chamber, the temperature control of the cavity, the magnetic field, etc.; and
2. to receive the signal emitted by the atoms and to use it in order to confer its stability and accuracy to the local oscillator.

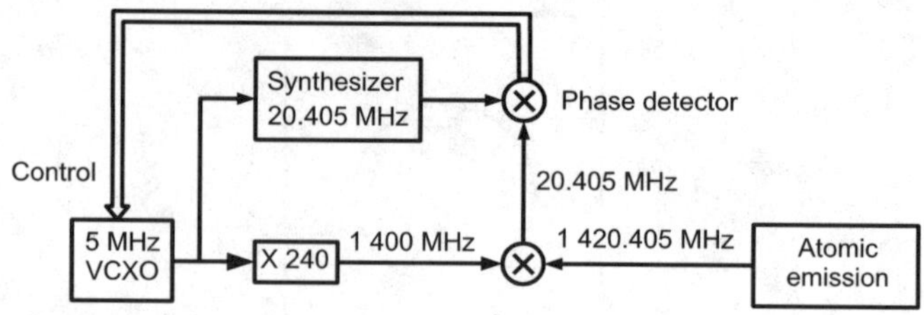

Fig. 2.16. Phase locking of the local oscillator on the atomic emission.

Contrary to the passive frequency standard (such as the cesium beam) where no atomic oscillation exists, it is possible to lock the phase of the local oscillator to that of the atomic radiation. This phase locking ensures the best values of the accuracy and stability of the local oscillator.

The electronic package is consequently made of

- a super heterodyne receiver that down converts the signal emitter by the atoms, transferring its frequency to a lower value by frequency mixing,
- a servo control of the phase of the local oscillator by the phase of the down-converted signal.

An example of this electronic package is shown in Fig. 2.16.

2.3.5 The Atomic Fountain

This kind of device [45, 69, 29], is quite similar to the cesium beam frequency standard described in Sect. 2.3.3. The main, and most important, difference

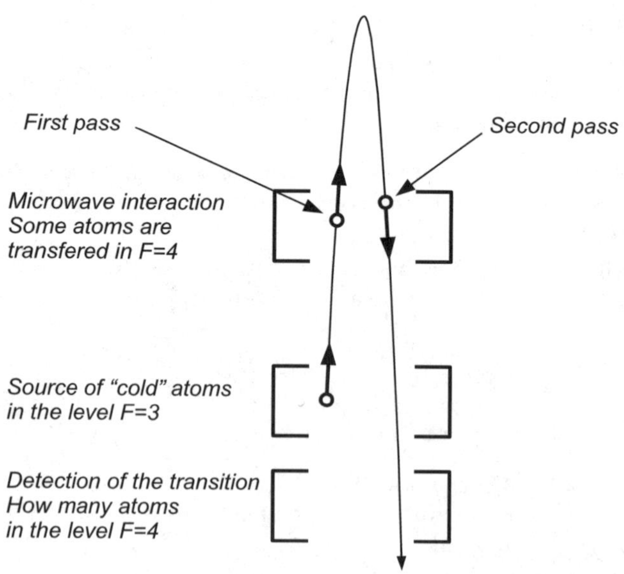

Fig. 2.17. Schematic diagram of an atomic fountain

lies in the fact that, in order to reduce the line width of the atomic transition, the mean velocity of the atoms in the cavity is greatly reduced by laser manipulation [103]. A cloud of very cold atoms is sent upward periodically, with a very low speed [70]. The trajectories of the atoms are parabolic and a given atom passes twice throw a single cavity: the first time upward and the second one downward, as shown in Fig. 2.17. As for the thermal beam, each atom consequently undergoes interaction with the probing transition twice.

Due to the very low mean velocity, the time interval T between the two interactions is much longer than in the case of a thermal beam and the line width is consequently reduced in the same proportion. Moreover, the velocity distribution of the atoms is greatly reduced, which simplifies the evaluation of the accuracy of the device.

The Physical Phenomenon

The schematic diagram of the detection of the hyperfine transition involving an atomic fountain of cesium is shown in Fig. 2.17.

Preparation of the Atoms

Atoms are cooled [103] by laser beams in a low pressure cesium vapor cell, producing optical molasses [82].

Clouds of atoms are periodically launched upwards with a velocity of a few $\mathrm{m} \times \mathrm{s}^{-1}$. Inside the moving cloud, the temperature of the atoms is lowered to a value of the order of 10^{-6} K. Atoms in the clock level $F = 3$ are selected by microwave and light pulses.

Probing of the Transition

Due to the low mean velocity of the atoms and the low velocity dispersion in each cloud, the mean value of the transition probability of the atoms that have passed twice in the probing cavity is

- much narrower: the full width of the central resonance is of the order of 1 Hz,
- very close of the transition probability of a monokinetic beam.

The calculated resonance curve is shown in Fig. 2.18 with the following values of the parameters:

- the mean value of the transit time in the cavity is $t = 1\,\mathrm{s}$,
- the time interval between the two passes in the cavity is $T = 20 \times t$,
- the most probable speed in the atomic cloud is 1.5×10^{-2}, the mean speed of the cloud when it passes throw the cavity,
- the Rabi frequency is $b = 10\,\mathrm{s}^{-1}$.

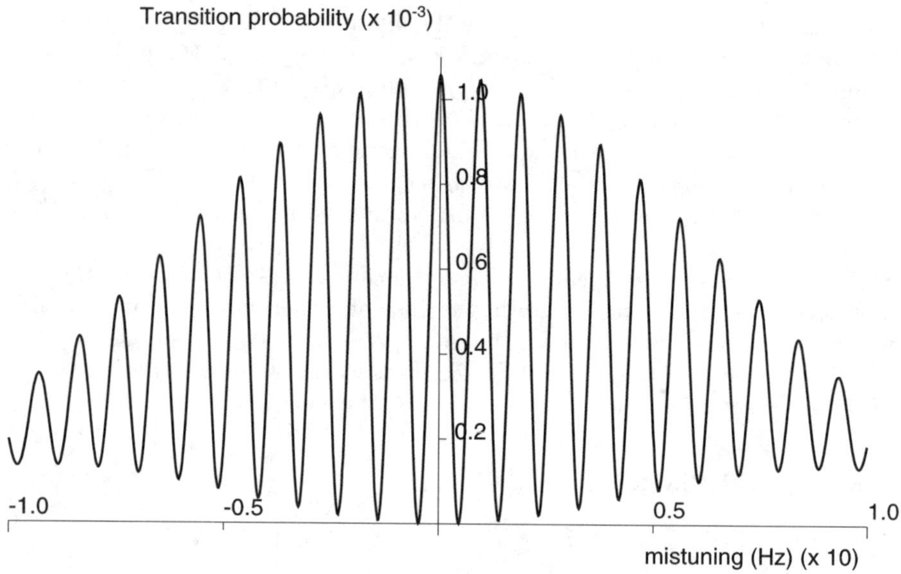

Fig. 2.18. Calculated Ramsay fringes in an atomic fountain

The velocity dispersion appears as a weak modulation of the fringes, instead of blurring them as in the case of a thermal atomic beam; many fringes are visible and one must pay attention to lock the local oscillator on the central one.

The experimental Ramsey fringes reported, for instance, in [30, 29, 125, 45, 69, 57] and [101] are quite similar to the calculated result of Fig. 2.18.

The Electronic Package

Like in the thermal cesium beam frequency standard, the functions of the electronic part of the cesium fountain frequency standard are as follows:

1. The control of the all physical experiment: cooling of the atoms, preparation of the atoms in one hyperfine level, launching of the atomic cloud.
2. The synthesis of the microwave excitation from the local oscillator.
3. The detection of the atoms that have undergone the transition.
4. The correction of the frequency of the local oscillator.

They are qualitatively identical to the functions of the electronic part of the thermal beam.

Nevertheless, an important difference results from the fact that the beam is pulsed: the atomic response is a discrete-time process, which leads to aliasing of the noise spectral density that must be taken into account (the Dick effect [42, 9, 84]).

This is discussed in Sect. 3.3.

3 Control of the Local Oscillator Frequency

It was pointed out in Chap. 2 that the quality of the oscillator depends on two kinds of factors:

1. those related to the physical phenomenon involved in the first part of the device, and
2. those related to the processing of the data given by the physical phenomenon in the second part of the device.

The first family is not within the scope of this book, but fixes the ultimate values of the metrological parameters of the device.

The second one must be optimized in order to obtain properties that are as close as possible to the ultimate ones.

As shown by the examples of Chap. 2, there are many different situations, but in each case

- a signal whose frequency f_1 is close to the frequency of the reference transition is generated from the local oscillator,
- an error signal results from a comparison of the frequency f_1 and the frequency of the clock transition, and
- this error signal allows the control of the frequency of the local oscillator.

Three representative situations will be studied and they represent the three examples of Sect. 2.3.

In the three cases, in order to optimize the properties of the atomic frequency standard, it is necessary to evaluate the frequency or/and phase noise of the controlled local oscillator, related to the possible sources of frequency or/and phase noise.

3.1 Active Frequency Standard

In this case (see Sects. 2.3.1 and 2.3.4), the atoms emit the reference signal and the phase of the local oscillator is controlled by the phase of this atomic signal.

Figure 2.16 shows an example of the phase control of the local oscillator by the atomic frequency in the case of the hydrogen maser.

3.1.1 The Phase Control Loop

We will focus our attention on a typical phase servo control loop such as the one shown in Fig. 2.16.

In such a loop:

1. a signal at a frequency close to the frequency of the atomic signal is coherently generated from the local oscillator to be controlled,
2. the phase difference between that signal and the atomic signal is converted to a voltage error signal) by a phase comparator, and
3. this voltage is processed by an electronic circuit (filter) and is applied to the control input of the local oscillator in order to minimize the phase difference.

The frequency delivered by the local oscillator is the superposition of

1. its nominal value f_{LO}, independent of time;
2. the correction produced by the voltage applied on its control input, a function of time; and
3. its frequency noise δf_{LO}, a function of time characterized by its Fourier transform $\delta f_{\mathrm{LO}}(\nu)$ and its spectral density (see Sect. 6.2.2).

The frequency delivered by the atoms is the superposition of

1. its theoretical frequency f_0, independent of time;
2. its frequency noise δf_0, a function of time characterized by its Fourier transform $\delta f_0(\nu)$ and its spectral density (see Sect. 6.2.2); and
3. the frequency shifts and frequency offsets δf_{sh} produced by the environment of the atoms.

The instantaneous frequency $f_{\mathrm{out}}(t)$ delivered by the device is the instantaneous frequency of the local oscillator.

Making the assumption that the instantaneous frequency of the signal produced by the frequency synthesis equals the instantaneous frequency of the local oscillator multiplied by the constant $n = f_0/f_{\mathrm{LO}}$, whose value is exactly fixed by the electronic device, the control loop can be represented as in Fig. 3.1.

In fact, the phase comparator delivers a voltage proportional to the phase difference between its two inputs + and −. The quantity of interest is consequently the phase rather than the frequency and the loop can be drawn with the phase as the input and output quantities (Fig. 3.2).

The block $1/\nu$ appearing in this figure corresponds to the integration operator that converts the frequency correction produced by the voltage applied on the frequency control input of the local oscillator to the phase correction of the signal delivered by this local oscillator. The phases ϕ_{LO} and ϕ_0 correspond to the nominal (and consequently constant) values of the frequencies delivered by the local oscillator and the atoms, respectively. The time-dependent parts are described by the noise sources and the offset and shifts source.

The gain of this multiple input loop can be calculated for each input.

The parameters of the components of the loop are as follows.

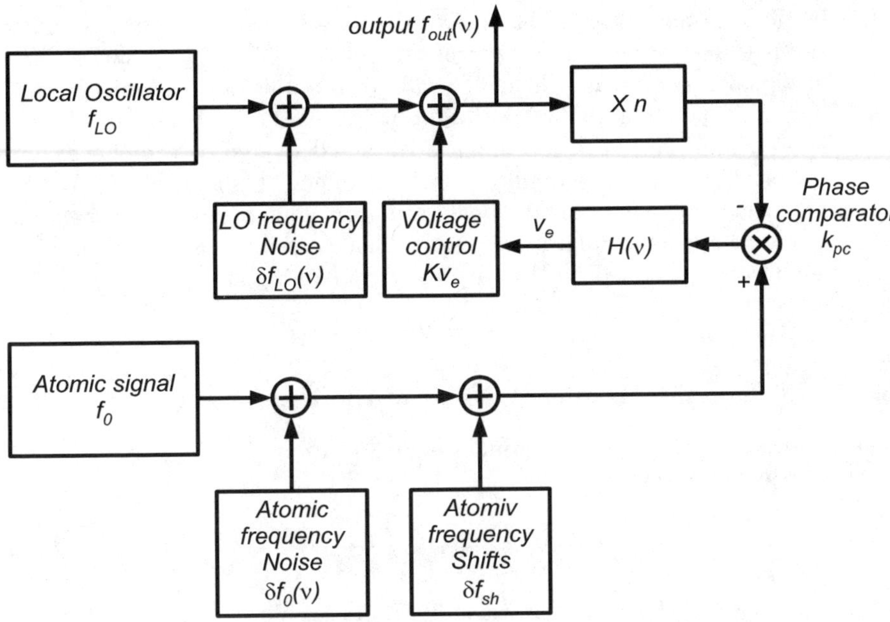

Fig. 3.1. Phase lock loop for the different components of the signals

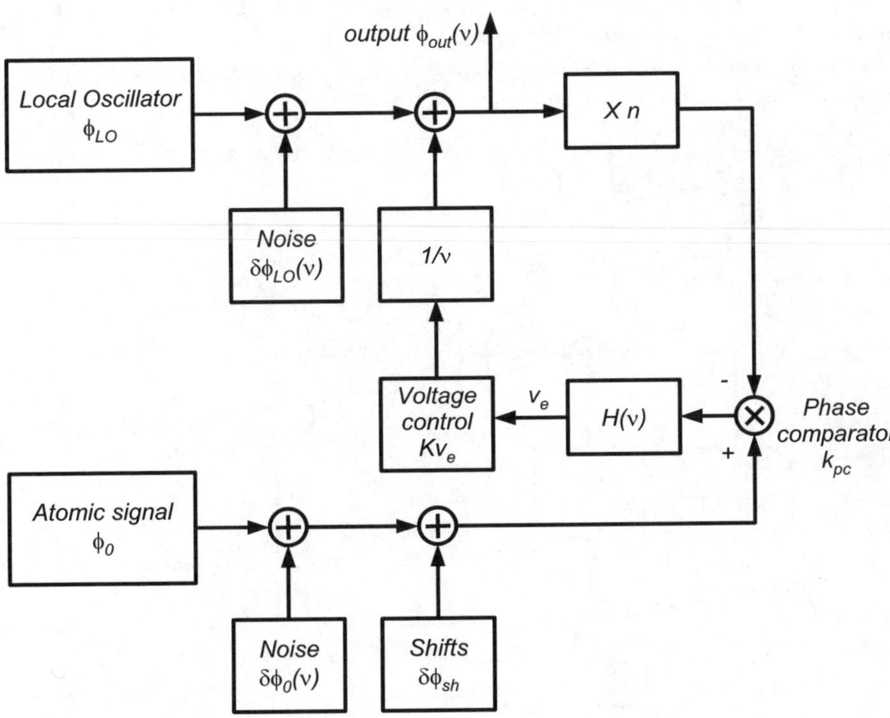

Fig. 3.2. The phase lock loop with the phase as the quantity of interest

1. The phase comparator. This receives the phases ϕ_+ and ϕ_- on its inputs $+$ and $-$, respectively, and returns the voltage $v_{\mathrm{pc}} = k_{\mathrm{pc}} \times (\phi_+ - \phi_-)$. The signal derived from the local oscillator is applied on its input ϕ_- and the atomic signal is applied on its input ϕ_+.
2. The electronic circuit processing the voltage delivered by the phase comparator. This is a low-pass filter whose frequency response is $H(\nu)$.
3. The frequency control of the local oscillator. This produces a correction δf_{con} to the local oscillator frequency. This correction is proportional to the voltage v_{e} applied on the frequency control input:

$$\delta f_{\mathrm{con}} = K \times v_{\mathrm{e}} \,.$$

3.1.2 Gain for the Frequency Noise of the Local Oscillator

The equivalent set-up for this input is shown in Fig. 3.3.

Its frequency response is

$$\frac{\phi_{\mathrm{out}}(\nu)}{\delta\phi_{\mathrm{LO}}(\nu)} = \frac{\nu}{\nu + nk_{\mathrm{pc}}KH(\nu)} \,. \tag{3.1}$$

In a general way, $H(\nu)$ is a low-pass filter, its amplitude response $G(\nu)$ is maximal and equals H_0 for the null Fourier frequency and is proportional to $1/\nu^o$, where o is the order of the filter, for values of the Fourier frequency much larger than the cut-off frequency of the filter ν_{c}.

Consequently, the amplitude response of the control loop for the phase noise of the local oscillator is

$$\left|\frac{\phi_{\mathrm{out}}(\nu)}{\delta\phi_{\mathrm{LO}}(\nu)}\right| = \begin{cases} \nu/(\nu + nk_{\mathrm{pc}}KH_0) & (\nu \ll \nu_{\mathrm{c}}) \\ 1 & (\nu \gg \nu_{\mathrm{c}}) \end{cases} \,. \tag{3.2}$$

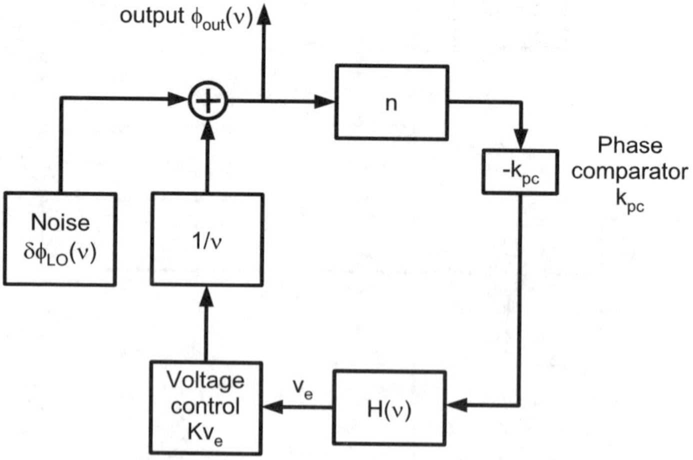

Fig. 3.3. Equivalent set-up for the local oscillator noise

In conclusion, the phase noise of the local oscillator is totally transmitted to the output of the device for values of the Fourier frequency much larger than the cut-off frequency of the filter and is strongly reduced for the values of the Fourier frequency much smaller than that cut-off frequency. It cancels exactly for $\nu = 0$.

3.1.3 Gain for the Frequency Noise of the Atomic Signal

The equivalent set-up for this input is shown in Fig. 3.4. Its frequency response is

$$\frac{\phi_{\text{out}}(\nu)}{\delta\phi_0(\nu)} = \frac{k_{\text{pc}}KH(\nu)}{\nu + nk_{\text{pc}}KH(\nu)} . \tag{3.3}$$

With the same filter as above, the asymptotic values of the amplitude response are the following:

$$\left|\frac{\phi_{\text{out}}(\nu)}{\delta\phi_0(\nu)}\right| = \begin{cases} \dfrac{k_{\text{pc}}KH_0}{\nu + nk_{\text{pc}}KH_0} \approx \dfrac{1}{n} & (\nu \ll \nu_{\text{c}}) \\ 0 & (\nu \gg \nu_{\text{c}}) \end{cases} . \tag{3.4}$$

In conclusion, the phase noise of the atomic signal is totally transmitted (divided by n, the ratio of the atomic signal and the local oscillator frequencies) to the output of the device for low values of the Fourier frequency. It is on the contrary totally blocked for the high values of the Fourier frequency.

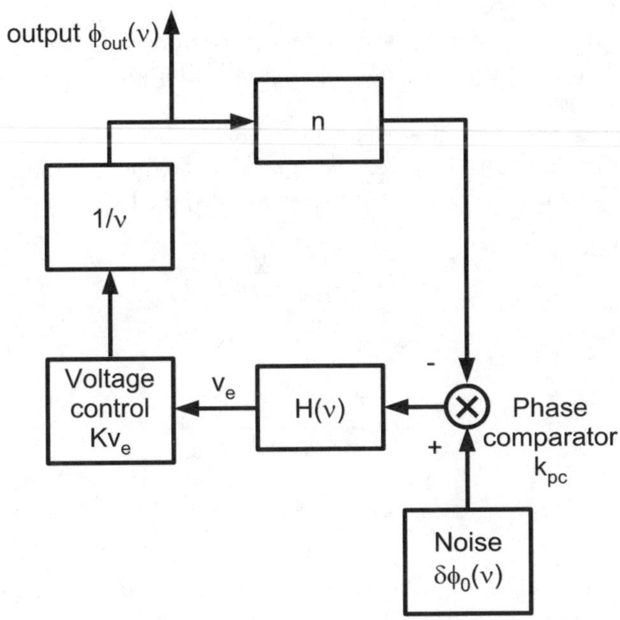

Fig. 3.4. Equivalent set-up for the noise of the atomic signal

3.1.4 Phase Noise of the Output

The phase noise (and consequently the frequency noise) of the signal delivered by the active atomic frequency standard is imposed either by the atomic emission or by the local oscillator, according to the Fourier frequency range as follows:

- the phase noise of the atomic signal is dominant in the output for the low values of the Fourier frequency, and
- the phase noise of the local oscillator is dominant in the output for large values of the Fourier frequency.

The Fourier frequency ν_0 for which the two contributions are equal is defined by the choice of the low-pass filter. The value of ν_0 is given as

$$\frac{\nu_0}{\nu_0 + nk_{\mathrm{pc}}K\,|H(\nu_0)|} \times |\delta\phi_{\mathrm{LO}}(\nu_0)| = \frac{k_{\mathrm{pc}}K\,|H(\nu_0)|}{\nu_0 + nk_{\mathrm{pc}}K\,|H(\nu_0)|} \times |\delta\phi_0(\nu_0)| \ , \tag{3.5}$$

$$k_{\mathrm{pc}}K\,|H(\nu_0)| = \nu_0 \frac{|\delta\phi_{\mathrm{LO}}(\nu_0)|}{|\delta\phi_0(\nu_0)|} \ . \tag{3.6}$$

3.1.5 Frequency Error

Apart from the effect of the noise sources, it is important to evaluate the static frequency error of the device, i.e. the difference between the output frequency (neglecting the effect of the various noise sources) and the atomic transition frequency.

This is done by calculating the frequency response of the loop at the null Fourier frequency for the nominal frequency of the local oscillator and that of the atomic signal.

This loop is shown in Fig. 3.5. The blocks with a transfer function ν and $1/\nu$ correspond to the conversions phase → frequency and frequency → phase. The following frequency responses are deduced from this loop:

$$\begin{aligned} \frac{f_{\mathrm{out}}}{f_0} &= \left.\frac{k_{\mathrm{pc}}KH(\nu)}{\nu + nk_{\mathrm{pc}}KH(\nu)}\right|_{\nu=0} \\ &= \frac{1}{n} \ , \end{aligned} \tag{3.7}$$

$$\begin{aligned} \frac{f_{\mathrm{out}}}{f_{\mathrm{LO}}} &= \left.\frac{\nu}{\nu + nk_{\mathrm{pc}}KH(\nu)}\right|_{\nu=0} \\ &= 0 \ . \end{aligned} \tag{3.8}$$

In conclusion

1. the frequency of the atomic oscillation is exactly transferred on the output of the device, and
2. the static nominal frequency of the local oscillator is totally rejected.

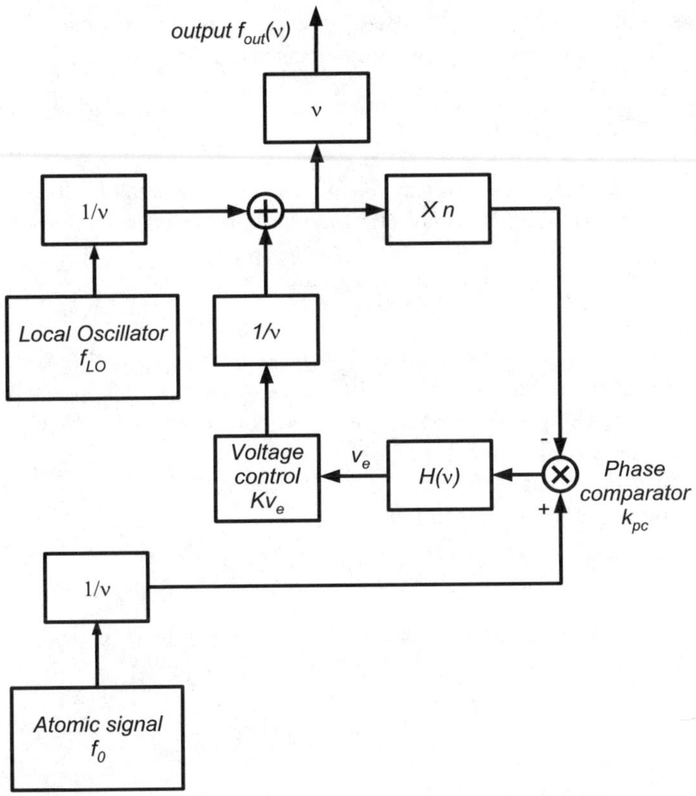

Fig. 3.5. The loop for the nominal frequency of the local oscillator and of the atomic oscillator

3.1.6 The Best Choices

The frequency responses for the different noise sources allow choosing the best components of the device to reduce as much as possible their contribution to the frequency noise of the output signal.

The Local Oscillator

1. The previous analysis shows that the short term stability of the active atomic frequency standard (which is limited by the frequency noise for the high values of the Fourier frequency) is fixed by the phase noise of the local oscillator. It is consequently necessary to choose a local oscillator with good short term stability.
2. It also shows that the stability of the atomic oscillation is transferred on the output signal of the device for the Fourier frequencies smaller than the value ν_0 defined by (3.6).

Consequently, for this reason also, it is necessary to choose a local oscillator with a low level of phase noise $\delta\phi_{\mathrm{LO}}(\nu)$.

The Phase Control Loop

For the small values of the Fourier frequency, the effect of the local oscillator phase noise on the phase noise of the output is reduced by the factor

$$\frac{\nu}{\nu + |nk_{\mathrm{pc}}KH_0|} \,. \tag{3.9}$$

It is consequently necessary to have the largest value possible for the zero-frequency amplitude response $|H_0|$ of the low-pass filter of the loop.

For instance, the phase noise of the good commercial 5 MHz quartz oscillator [98] is given in Table 3.1[1].

The phase noise spectral density of this quartz oscillator for Fourier frequencies close to 1 Hz is consequently

$$p_{\delta\phi}^{\mathrm{Q}}(\nu) = 6.32 \times 10^{-12} \times \nu^{-2}\,\mathrm{Hz}^{-1} \,. \tag{3.10}$$

The frequency noise spectral density is consequently (see (6.101)):

$$p_{\delta f}^{\mathrm{Q}}(\nu) = \nu^2 \times p_{\delta\phi}^{\mathrm{Q}}(\nu) \tag{3.11}$$

$$= 6.32 \times 10^{-12}\,\mathrm{Hz} \,. \tag{3.12}$$

This is a white frequency noise.

The reduced frequency noise spectral density of the quartz oscillator is

$$p_y^{\mathrm{Q}} = p_{\delta f}^{\mathrm{Q}}(\nu)/f_0^{\mathrm{Q}\,2} \tag{3.13}$$

$$= 2.53 \times 10^{-25}\,\mathrm{Hz}^{-1} \,. \tag{3.14}$$

This corresponds to the reduced white frequency noise $p_y^{\mathrm{Q}}(\nu) = h_0^{\mathrm{Q}}$ with $h_0^{\mathrm{Q}} = 2.53 \times 10^{-25}\,\mathrm{Hz}^{-1}$ (see Table 6.1).

On the other hand, the typical value of the Allan variance of a hydrogen maser for sampling times around 1 s is

$$\sigma_y^2(\tau) = 1 \times 10^{-26} \times \tau^{-2} \,, \tag{3.15}$$

which corresponds to a reduced frequency noise $p_y^{\mathrm{H}}(\nu) = h_2^{\mathrm{H}}\nu^2$ with $h_2^{\mathrm{H}} \approx 10^{-26}\,\mathrm{Hz}^{-1}$.

In order to impose, for instance, that the Fourier frequency for which the spectral density of the frequency noise of the local oscillator and that of the

[1] The values are given in dBc, defined by $\mathcal{L}(\nu) = 10\log[0.5(S_\phi(\nu))]$ where $S_\phi(\nu)$ is the phase noise spectral density in units of Hz^{-1}.

Table 3.1. Phase noise of a commercial quartz oscillator

Fourier frequency (Hz)	Spectral density (dBc)	Spectral density (Hz^{-1})
1	−115	6.32×10^{-12}
10	−135	6.32×10^{-14}
100	−145	6.32×10^{-15}
1 000	−150	2.00×10^{-12}

spectral density of the frequency noise of the atomic oscillation give the same contribution to the output of the device be 1 Hz, application of (3.6) gives

$$k_{pc} K \left| H(\nu) \right|_{\nu = 1\,\mathrm{Hz}} = \left. \frac{|\delta\phi_{LO}(\nu)|}{|\delta\phi_0(\nu)|} \right|_{\nu = 1\,\mathrm{Hz}} = 25.3\,. \tag{3.16}$$

If k_{pc} (the response of the phase comparator) and K (the response of the voltage control of the local oscillator) are given, this relation imposes a minimal value for the amplitude response of the low-pass filter at the 1 Hz frequency.

3.2 Passive Frequency Standard

In this case (see Sect. 2.3.1), the atomic collection is too dilute to emit the reference signal. The clock transition is probed by a signal generated from the local oscillator and an error signal results from the atomic response. This error signal allows control of the frequency of the local oscillator.

Consequently, the frequencies *but not the phases* of the local oscillator and of the atomic transition are compared; the integration due to the conversion frequency → phase does not appear in the loop and its performances around the null Fourier frequency are consequently less good than in the case of the active frequency standard.

Moreover, the frequency comparison is sensitive to the amplitude fluctuations of the atomic response, and consequently to the amplitude noise affecting this signal. This noise must consequently be taken into account.

Figure 2.11 shows the simplified block diagram of an electronic device that synthesizes the frequency f_1, close to the frequency f_{Cs} of the clock transition of the cesium atom [90].

The principle of the control loop is shown in Fig 3.6: the response of the atoms to the frequency modulated excitation is lock-in detected. The lock-in amplifier produces a voltage proportional to the difference $f_1 - f_{Cs}$ of the frequency f_1 of the probing signal and the frequency f_{Cs} of the atomic transition. This error signal is used to control the frequency of the local oscillator.

S is the transfer function of this lock-in amplifier (V/Hz).

The sources of frequency noise implied in the frequency-control loop are shown in Fig. 3.7. They are:

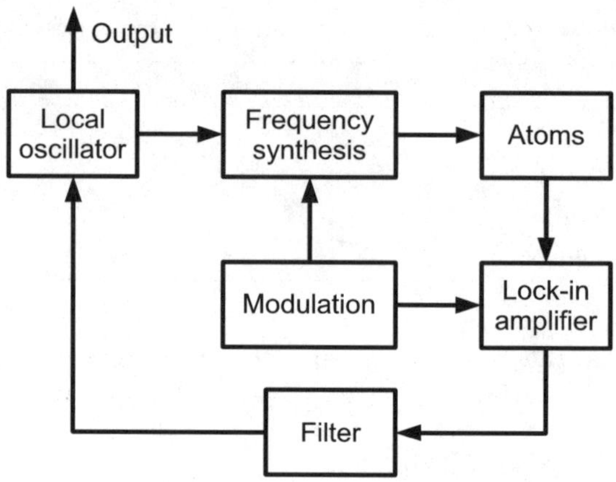

Fig. 3.6. Control of the frequency of the local oscillator in a passive atomic frequency standard

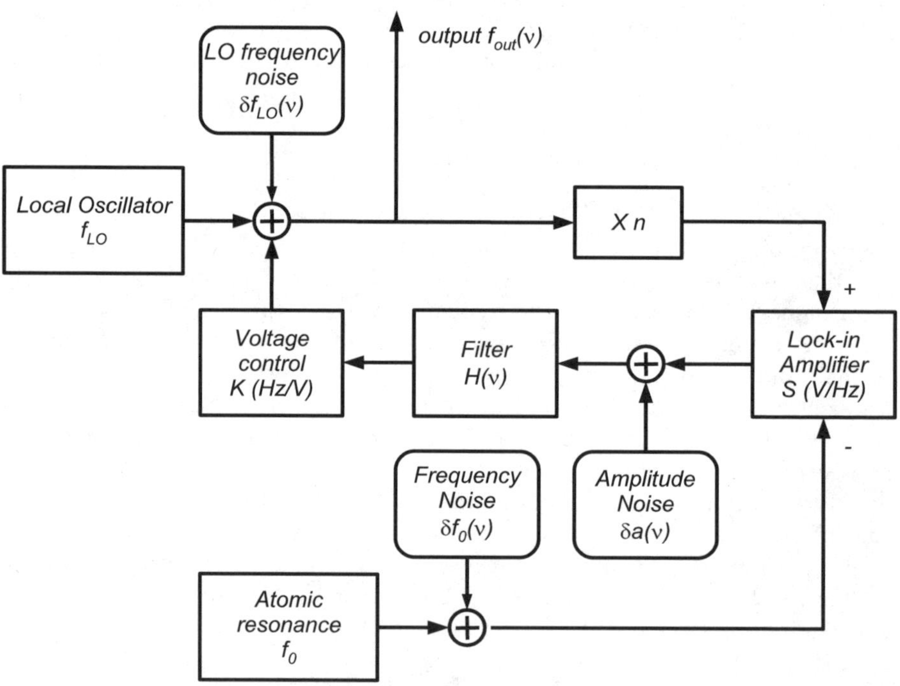

Fig. 3.7. Model for the noise in the frequency control loop

- the intrinsic frequency noise of the local oscillator reflected on the excitation signal as described in the case of the active atomic frequency standard (Sect. 3.1.1);
- the frequency noise of the atomic resonance; and
- the amplitude noise of the atomic response.

3.2.1 Frequency Response for the Frequency Noise of the Local Oscillator

The frequency response of the loop for the frequency noise of the local oscillator is

$$\frac{f_{\mathrm{out}}(\nu)}{\delta f_{\mathrm{LO}}(\nu)} = \frac{1}{1 - nSKH(\nu)} \; . \tag{3.17}$$

The sign of the product nSK must, of course, be selected suitably in order to ensure the stability of the loop.

As for the active atomic frequency standard (Sect. 3.1), $H(\nu)$ is the frequency response of a low-pass filter. The amplitude response $|H(\nu)|$ is maximal and equals H_0 for the null Fourier frequency and is proportional to $1/\nu^o$, where o is the order of the filter, for values of the Fourier frequency much larger than the cut-off frequency of the filter ν_c.

The value $|H_0|$ may be very large.

The amplitude response of the control loop for the frequency noise of the local oscillator is

$$\left| \frac{f_{\mathrm{out}}(\nu)}{\delta f_{\mathrm{LO}}(\nu)} \right| = \begin{cases} |1/(1 - nSKH_0)| & (\nu \ll \nu_c) \\ 1 & (\nu \gg \nu_c) \end{cases} \; . \tag{3.18}$$

The conclusion is qualitatively the same as for the active atomic frequency standard: the frequency noise of the local oscillator is totally transmitted to the output of the device for values of the Fourier frequency much larger than the cut-off frequency of the filter and is strongly reduced for the values of the Fourier frequency much smaller than that cut-off frequency.

Nevertheless, it is to be noticed that *the contribution of the local oscillator to the output frequency noise for the low values of the Fourier frequency is not totally rejected.*

3.2.2 Frequency Response for the Frequency Noise of the Atomic Resonance

The frequency response of the loop for the frequency noise of the atomic resonance is

$$\frac{f_{\mathrm{out}}(\nu)}{\delta f_0(\nu)} = \frac{-SKH(\nu)}{1 - nSKH(\nu)} \; . \tag{3.19}$$

With the same filter as above, the limits are the following for the amplitude response:

$$\left|\frac{f_{\mathrm{out}}(\nu)}{\delta f_0(\nu)}\right| = \begin{cases} |SKH_0/(1-nSKH_0)| \approx 1/n & (\nu \ll \nu_{\mathrm{c}}) \\ 0 & (\nu \gg \nu_{\mathrm{c}}) \end{cases} . \tag{3.20}$$

The conclusion is qualitatively the same as for the active atomic frequency standard: the frequency noise of the atomic signal is totally transmitted (divided by n, the ratio of the atomic signal and the local oscillator frequencies) to the output of the device for low values of the Fourier frequency. On the other hand, it is totally blocked for high values of the Fourier frequency.

3.2.3 Frequency Response for the Amplitude Noise of the Atomic Resonance

This contribution must be considered because amplitude fluctuations of the atomic signal are directly transmitted to the output of the lock-in amplifier. The frequency response of the loop for this input is

$$\frac{f_{\mathrm{out}}(\nu)}{\delta a(\nu)} = \frac{KH(\nu)}{1-nSKH(\nu)} . \tag{3.21}$$

The effect of amplitude noise $a(\nu)$ of the atomic response is consequently the same as a frequency noise $\delta f_{\mathrm{a}}(t)$ affecting the atomic frequency. The Fourier transform of this frequency noise is defined by

$$\delta f_{\mathrm{a}}(\nu) = -\frac{1}{S} \times \delta a(\nu) . \tag{3.22}$$

This frequency noise is totally transmitted on the output of the device (divided by n, the ratio of the atomic signal and the local oscillator frequencies) and is often the main limitation on the frequency stability.

3.2.4 Limitation to the Stability Due to the Amplitude Noise of the Atomic Response

Assuming a white amplitude noise affecting the atomic response,

$$p_{\mathrm{a}}(\nu) = a_0 . \tag{3.23}$$

The spectral density $p^{\mathrm{a}}_{\mathrm{out}}(\nu)$ of the resulting frequency noise for the low Fourier frequencies is

$$p^{\mathrm{a}}_{\mathrm{out}}(\nu) = p_{\mathrm{a}}(\nu) \times \frac{1}{n^2S^2} \tag{3.24}$$

$$= \frac{a_0}{n^2S^2} . \tag{3.25}$$

If this contribution is the dominant one, the resulting Allan variance is (see (6.182))

$$\sigma_y^2(\tau) = \frac{a_0}{2n^2S^2} \times \frac{n^2}{f_0^2} \times \tau^{-1}$$
$$= \frac{a_0}{2f_0^2S^2} \times \tau^{-1} . \tag{3.26}$$

On the other hand, the transfer function S (the slope of the discriminator equivalent to the resonance probing) can be calculated, knowing the detail of the probing process (see, for instance [90, 112]).

In all cases, the slope S of the frequency discriminator is

- proportional to the height $\sin^2(bt)$ of the central fringe of the Ramsey pattern, as defined in (2.6); and
- inversely proportional to the width $1/2T$ of this fringe, as defined in (2.7).

The Rabi frequency b is proportional to the amplitude of the magnetic field in the cavity

$$b = \frac{\mu_\mathrm{B}}{\hbar} , \tag{3.27}$$

where t is the interaction time of the atoms with the excitation in each region of the cavity and T is the transit time of the atoms between the two interaction regions of the cavity. The Allan variance corresponding to this frequency noise of the output signal (whose frequency is f_0/n) is

$$\sigma_y^2(\tau) = \frac{a_0}{8T^2f_0^2\sin^4(bt)} \times \tau^{-1} . \tag{3.28}$$

This relation shows the importance of the optimization of the product bt and of the value of the transit time T, which must be as large as possible.

The value of this contribution to the Allan variance of the device output signal varies like τ^{-1}: when τ becomes larger and larger, it is masked by another contribution that decreases less quickly with τ.

The first contribution to be considered is one that should give an Allan variance independent of τ (τ^0).

Table 6.3 shows that it corresponds to a frequency noise whose spectral density is $h_{-1}\left|\nu^{-1}\right|$ (flicker noise). Flicker noise is the long-term limitation of cesium frequency standards.

Table 3.2. Guaranteed values of the Allan deviation of a commercially available cesium frequency standard

Averaging time τ (s)	Allan deviation
$10^1 \leq \tau \leq 4.32 \times 10^5$	$2.7 \times 10^{-11}/\sqrt{\tau}$
$4.32 \times 10^5 \leq \tau \leq 2.59 \times 10^6$	5.0×10^{-14}

For instance, the commercial frequency standard [2, 113] shows some guaranteed values of Allan variance (see Table 3.2). These values show the contribution of the white amplitude noise affecting the atomic signal and the limit due to flicker noise (flicker floor).

Another example is given by the primary frequency standards [90] whose Allan deviation is $\sigma_y(\tau) = 3.5 \times 10^{-13}/\sqrt{\tau}$ for $60\,\mathrm{s} \leq \tau \leq 1 \times 10^5\,\mathrm{s}$.

3.2.5 Frequency Error

Neglecting the noise sources, it is important to evaluate the static frequency error of the device, i.e. the difference between the output frequency (supposedly not affected by the various noise sources) and the atomic transition frequency. This is done by calculating the frequency response of the loop at the null Fourier frequency for the nominal frequency of the local oscillator and for that of the atomic signal.

This loop is shown in Fig. 3.8. The following frequency responses result:

$$\begin{aligned} \frac{f_{\mathrm{out}}}{f_0} &= \left.\frac{SKH(\nu)}{1+nSKH(\nu)}\right|_{\nu=0} \\ &\approx \frac{1}{n}\,, \end{aligned} \tag{3.29}$$

$$\begin{aligned} \frac{f_{\mathrm{out}}}{f_{\mathrm{LO}}} &= \left.\frac{1}{1+nSKH(\nu)}\right|_{\nu=0} \\ &\approx \frac{1}{nSKH_0}\,. \end{aligned} \tag{3.30}$$

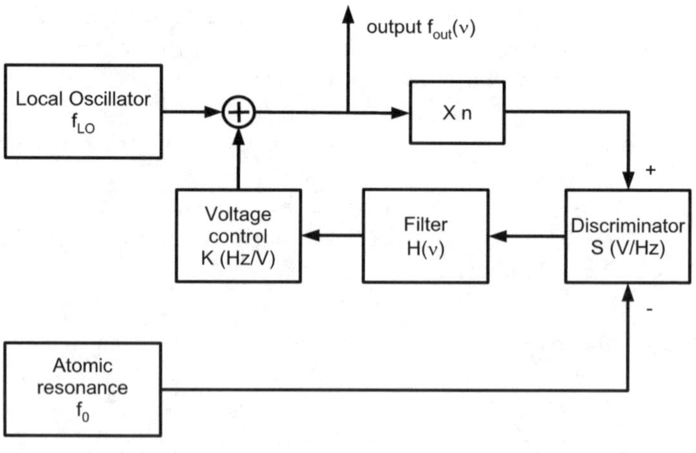

Fig. 3.8. The loop for the nominal frequency of the local oscillator and of the atomic transition

In conclusion:

1. The frequency of the atomic oscillation is *not* exactly transferred on the output of the device. The relative error is
$$\left|\frac{f_{\mathrm{out}} - f_0/n}{f_0/n}\right| = \frac{1}{|SKH_0|} . \tag{3.31}$$
This error is smaller if the static gain of the low-pass filter is higher.
2. The static nominal frequency of the local oscillator is *not* totally rejected. The attenuation factor is $nSKH_0$. It is larger if the static gain of the low-pass filter is higher.

These conclusions show the need to obtain the static gain of the low-pass filter as high as possible.

3.2.6 The Best Choices

The conclusions relative to the design of the loop are qualitatively the same as in the case of the active atomic frequency standard.

The Local Oscillator

1. The short term stability of the active atomic frequency reference is limited by the phase noise of the local oscillator. It is consequently necessary to choose a local oscillator with good short-term stability.
2. Assuming that the amplitude noise affecting the atomic response is the limitation of the dominant process, the stability of the atomic oscillation is transferred to the output signal of the device for Fourier frequencies smaller than the value ν_0 defined by
$$|a(\nu_0)KH(\nu_0)| = |\delta f_{\mathrm{LO}}(\nu_0)| . \tag{3.32}$$
In order to obtain the largest value of this frequency, the frequency noise level of the local oscillator must be as low as possible.

The Phase Control Loop

The effect of the local oscillator frequency noise on small values of the Fourier frequency is reduced by the factor
$$\frac{1}{|1 + nSKH_0|} . \tag{3.33}$$

It is consequently necessary to have the largest value possible for the zero-frequency amplitude response $|H_0|$ of the low-pass filter of the loop.

3.3 The Sampled Servo-loop

Some passive atomic frequency standards involve a sampled-and-hold error signal.

1. If the modulation of the excitation frequency is a square wave, the transient atomic response occurring during a first part of each half period is ignored and the permanent state response is integrated during the remaining of each half-period of modulation. Two successive results are used to construct the value of the error signal at times nT_l, where T_l is the period of the square wave modulation.
2. Some devices such as atomic fountains (see Sect. 2.3.5) or stored ions bring into play a temporal sequence: preparation – excitation – interrogation. The response is consequently a discrete-time signal whose value is known only at the end of each temporal sequence of duration T_l.

In this family of devices, aliasing of the noise spectra may be produced by the sample-and-hold operator (the Dick effect [42, 9, 84]) and must be taken into account.

3.3.1 Model of the Servo-loop in the Case of an Atomic Fountain

This kind of atomic frequency standard is described in Sect. 2.3.5.

We call T_l the sampling interval, i.e. the time interval between two successive launchings of cold atoms by the atomic source.

The local oscillator is free running between two successive sampling times (the voltage frequency control of the local oscillator is hold) and corrections are updated at the times nT_l, i.e. the times when an atomic cloud is detected at the end of a cycle preparation – excitation – interrogation.

Figure 3.9 shows the frequency lock loop of Fig. 3.7 where the error signal delivered by the frequency discriminator is sampled and held at times nT_l.

The sampler and hold is an operator that strongly affects the Fourier transform (and consequently the spectral density) of any process whose frequency range extends beyond the Nyquist frequency (which is half of the sampling frequency $1/2T_l$).

Assuming a continuous-time process $f(t)$ whose Fourier transform is $F(\nu)$, the Fourier transform $F'(\nu)$ of the continuous-time process resulting from sampling-and-hold $f(t)$ is (see (B.63))

$$F'(\nu) = \exp(-j\pi\nu T_l) \times \mathrm{sinc}(\pi\nu T_l) \times \sum_{k=-\infty}^{\infty} F(\nu + k/T_l) \; . \tag{3.34}$$

The Fourier component of frequency $\nu > 1/2T_l$ is consequently modified by this process.

In order for this effect to be handled easily, the loop can be described by two simplified models, each one being applied in a different Fourier frequency range.

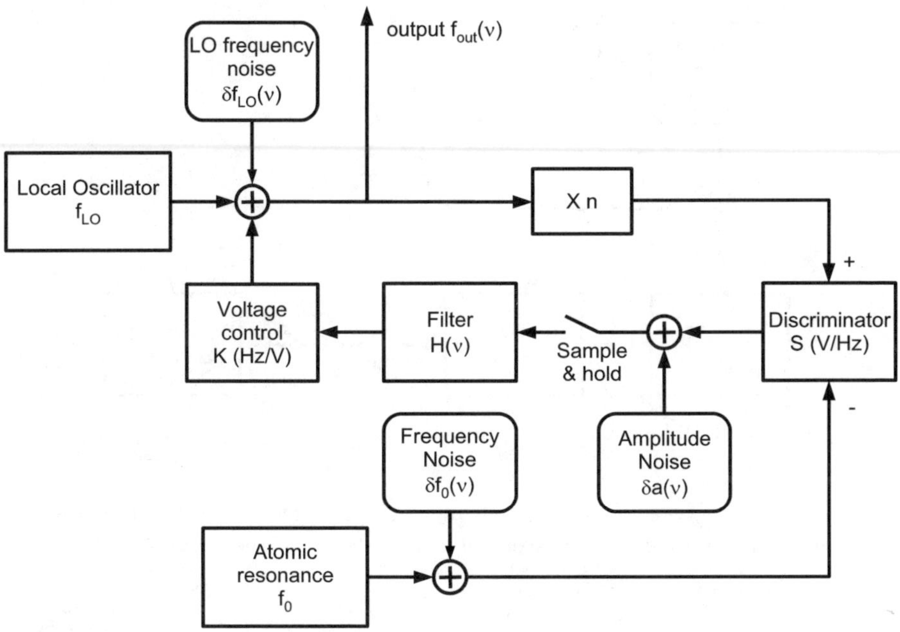

Fig. 3.9. Sampling in the loop of Fig. 3.7

1. Model for the high values of the Fourier frequency, which are strongly attenuated by the low-pass filter. These frequencies consequently exist only between every source that produces them in the set-up described in Fig. 3.9 and the low-pass filter, and also between the sampling-and-hold and the low-pass filter, since this device produces high frequency components.
2. Model for the low values of the Fourier frequency. These frequencies exist in the whole set-up of Fig. 3.9. The low-pass filter is such that this frequency range is supposed to be inside the range $[-1/T_{\text{l}}, 1/T_{\text{l}}]$ and is not modified by the sampling-and-hold operation.

3.3.2 Frequency Responses for the Frequency Noise of the Local Oscillator

The loop for the local oscillator frequency noise is shown in Fig. 3.10.

In the simplified model described in Sect. 3.3.1, the noise emitted by the local oscillator frequency noise source, not yet filtered by the low-pass filter, may contain frequency components larger than the Nyquist frequency. Consequently, frequency aliasing of the frequency noise of the local oscillator occurs and must be taken into account.

On the other hand, the node f_1 (Fig. 3.10) contains no Fourier component outside the Nyquist range. The Fourier components outside the Nyquist range

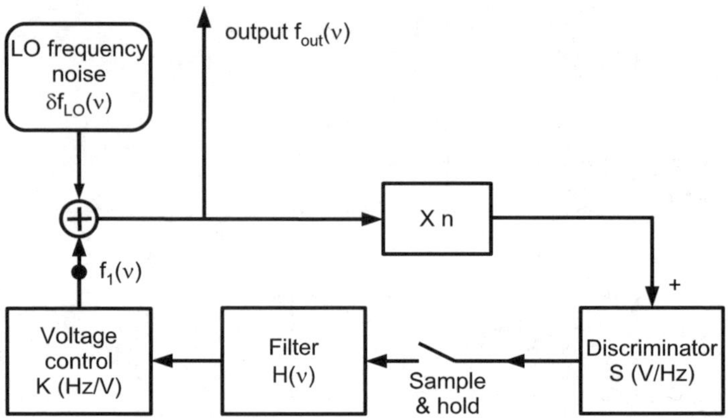

Fig. 3.10. The sampled loop for the local oscillator frequency noise

on node f_{out} are consequently only due to the frequency noise source and the following equation can be written for the nodes f_{out} and f_1.

1. For the low frequency range

$$f_{\text{out}}(\nu) = \delta f_{\text{LO}}(\nu) + f_1(\nu) , \tag{3.35}$$

$$f_1(\nu) = -G_{\text{OL}}(\nu) \left[f_{\text{out}}(\nu) + \sum_{k \neq 0} \delta f_{\text{LO}}(\nu + k/T_{\text{l}}) \right] . \tag{3.36}$$

2. For the high frequency range

$$f_{\text{out}}(\nu) = \delta f_{\text{LO}}(\nu) , \tag{3.37}$$

$$f_1(\nu) = 0 . \tag{3.38}$$

$G_{\text{OL}}(\nu)$ is the loop gain

$$G_{\text{OL}}(\nu) = -nSH(\nu)K\text{sinc}(\pi\nu T_{\text{l}})\exp(-j\pi\nu T_{\text{l}}) . \tag{3.39}$$

In conclusion, for very low values of the Fourier frequency ν (such that $|G_{\text{OL}}(\nu)| \gg 1$)), the direct contribution of the local oscillator frequency noise vanishes, but the effect of frequency aliasing due to the sampling brings the contribution

$$f_{\text{out}}(\nu) \approx \sum_{k \neq 0} \delta f_{\text{LO}}(\nu + k/T_{\text{l}}) . \tag{3.40}$$

Special care must be taken to reduce this contribution by choosing the local oscillator. This local oscillator is often a hydrogen maser (see [57] for an example). On the other hand, for large values of the Fourier frequency, (3.37) shows that the frequency noise of the local oscillator is totally transmitted on the output of the device.

3.3.3 Frequency Response for the Amplitude Noise of the Atomic Transition

In the simplified model (see Sect. 3.3.1), the loop can be described according to the Fourier frequency range by (see Fig. 3.11) one of the two following sets of equations:

1. For the low frequency range

$$f_{\text{out}} = \left[\delta a(\nu) + f_1(\nu) + \sum_{k \neq 0} \delta a(\nu + k/T_1) \right] \times KH(\nu) \text{sinc}(\pi\nu T_1) \exp(-j\pi\nu T_1) \; , \tag{3.41}$$

$$f_1(\nu) = f_{\text{out}}(\nu) nS \; , \tag{3.42}$$

which gives

$$f_{\text{out}}(\nu) = \frac{KH(\nu)\text{sinc}(\pi\nu T_1)\exp(-j\pi\nu T_1)}{1 + G_{\text{OL}}(\nu)} \left[\sum_{k=-\infty}^{\infty} \delta a(\nu + k/T_1) \right] . \tag{3.43}$$

$G_{\text{OL}}(\nu)$ is the loop gain

$$G_{\text{OL}}(\nu) = -nSH(\nu)KT_1 \text{sinc}(\pi\nu T_1) \exp(-j\pi\nu T_1) \; . \tag{3.44}$$

2. For the high frequency range

$$f_{\text{out}}(\nu) = 0 \; . \tag{3.45}$$

In conclusion, the effect of the amplitude noise of the atomic response cancels for high values of the Fourier frequency, as is the case for a continuous-time loop.

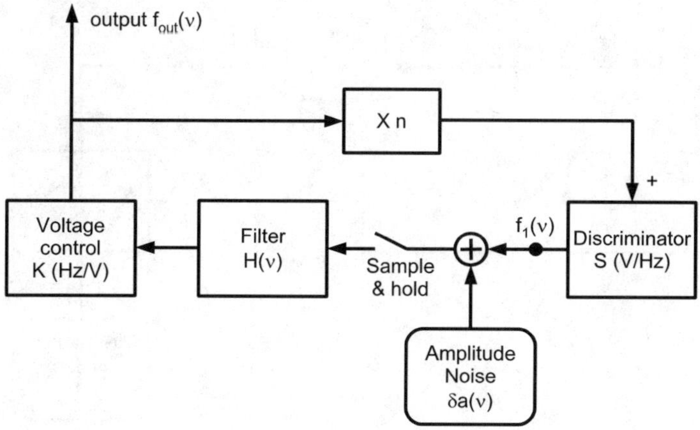

Fig. 3.11. The sampled loop for the amplitude noise

On the other hand, the effect of the amplitude noise, which is dominant for low values of the Fourier frequency (long-term stability), is amplified by aliasing as shown in (3.43).

With the same simplified model applied to the set-up of Fig. 3.12, the following equations can be written:

1. For the low frequency range

$$f_{\text{out}} = \left[-S\delta f_0(\nu) + f_1(\nu) - S \sum_{k \neq 0} \delta f_0(\nu + k/T_1) \right] \times KH(\nu)\text{sinc}(\pi\nu T_1) \exp(-j\pi\nu T_1) , \tag{3.46}$$

$$f_1(\nu) = f_{\text{out}}(\nu) nS , \tag{3.47}$$

which gives

$$f_{\text{out}}(\nu) = \frac{-SKH(\nu)\text{sinc}(\pi\nu T_1) \exp(-j\pi\nu T_1)}{1 + G_{\text{OL}}(\nu)} \times \left[\sum_{k=-\infty}^{\infty} \delta f_0(\nu + k/T_1) \right] , \tag{3.48}$$

$G_{\text{OL}}(\nu)$ is the loop gain

$$G_{\text{OL}}(\nu) = -nSH(\nu)KT_1\text{sinc}(\pi\nu T_1) \exp(-j\pi\nu T_1) . \tag{3.49}$$

2. For the high frequency range

$$f_{\text{out}}(\nu) = 0 . \tag{3.50}$$

In conclusion, the effect of the frequency noise is qualitatively the same as the effect of the amplitude noise, with the factor $-S$: it cancels for the high frequency range and is worsened by aliasing for the low frequency range.

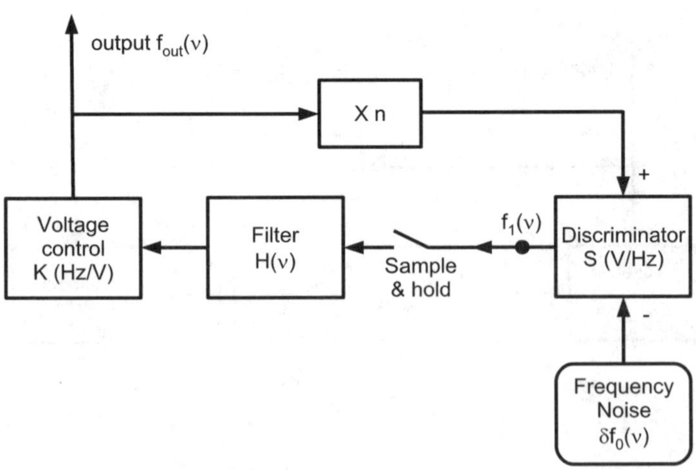

Fig. 3.12. The sampled loop for the frequency noise of the atomic transition

Part II

Characterization of Very High Performances Oscillators

High performance frequency sources allow one to achieve measurements leading to a best knowledge of our world, from a very concrete and practical point of view as well as a fundamental one.

If a given application involves a specified high performance frequency source, it is important to decide the following:

- whether the specification can be achieved using an existing frequency source, and
- what kind of frequency source is the best choice to fulfill the objective.

This first stage in the design of the application implies a perfect knowlege of the performances and limitations of the frequency sources.

These performances are characterized by a few parameters whose signification must be well understood. They are as follows:

- stability, with the indication of the time duration on which the given value is measured,
- accuracy,
- reproducibility, and
- sensitivity to the environment.

These parameters are often described as the "metrological properties" of the frequency source.

The following parameters, which describe the practical use of the device, must of course also be taken into account:

- the price,
- the weight and physical dimensions,
- the power consumption,
- the reliability, and
- the ability to run without any human intervention during a given time interval, etc.

Only the parameters of the first group need to be precisely described and discussed.

The following chapters describe the accuracy, the reproducibility and the stability of high performance frequency sources. The concepts and the meaning of these parameters are described, as well as their measurements.

4 Accuracy

4.1 Definition

According to the definition given by NIST [96], Accuracy is *the degree of conformity of a measured or calculated value to its definition.*

The accuracy of a device that produces a frequency characterizes the offset of this frequency vs. the theoretical (nominal) one.

For instance, the accuracy of a 10 MHz frequency source is defined by the difference between the value of the actual frequency of the output signal delivered by this device (for instance 9.9995 MHZ) and the nominal one (10 MHz).

It is often expressed by a positive dimensionless number, which is the relative offset (the absolute value of the possible frequency offset divided by the nominal value).

For the 10 MHz oscillator whose real frequency is, for instance, 9.9995 MHz, the accuracy is

$$Ac = \frac{|10 - 9.9995|}{10} = 5 \times 10^{-4} \, .$$

The accuracy of a device that produces a time-scale is the difference between a measured on-time pulse and an ideal on-time pulse that coincides exactly with UTC.

This accuracy is expressed in seconds: if the time difference between the pulse produced by the time-scale device and UTC is one microsecond, the accuracy equals 1 μs.

Notice that the smaller the number giving the accuracy, the better the device. For this reason, the term *uncertainty* is most often used when a quantitative measure is stated. Accuracy is used in a qualitative sense.

In fact, uncertainty is a statistical data, which characterizes the statistical result of the various errors affecting the value of the frequency delivered by the device. It is quite different from an offset, which may be easily corrected.

4.2 Estimation of the Uncertainty of a Frequency Source

In order to evaluate the uncertainty of the frequency emitted by a given device, one must do the following [64, 63]:

1. Identify the components contributing to the uncertainty, i.e. the phenomena that produce a departure of the frequency of the oscillator from its nominal value.
2. Quantify these individual components. This quantification can be performed according to two types of methods.
   - The type A evaluation of uncertainty, which uses statistical methods (statistical uncertainties).
   - The type B evaluation of uncertainty, using non-statistical methods (systematic uncertainties). In the case of frequency standards, type B uncertainties are evaluated by calculating or measuring the frequency offset produced by the phenomenon that perturbs the oscillator frequency and the uncertainty which affects this calculation or measurement.
3. Combine these components in order to quantitatively characterize the overall uncertainty of the frequency. This is the quadrature sum of the uncertainties.

This estimation is called the uncertainty budget of the frequency standard.

As an example, consider the uncertainty budget of the primary frequency standard NIST-F1, as described in [58] and [57].

The physical effects that are taken into account are briefly described below. For each of them, the magnitude and uncertainty of the frequency bias it produces are studied and the results are given.

1. Gravitational redshift (type B). This relativistic effect is discussed in Sect. 7.1.1. Its precise evaluation depends on the knowledge of the position of the frequency standard compared to the geoid. The value of the fractional frequency bias reported in [57] is 180.54×10^{-15} and an uncertainty on this value of 0.03×10^{-15}.
2. The second-order Zeeman effect (type B). This effect is the frequency bias produced by the weak and well-controlled magnetic field applied on the atoms [139, 34, 136]. In the case of [57], the fractional frequency bias is 36.53×10^{-15} and its uncertainty is 0.02×10^{-15}.
3. Blackbody radiation (type B). The perturbation created by the oscillatory electrical field due to black body radiation at the cavity temperature T shifts the atomic energy levels [91]. The fractional frequency bias calculated in [57] is -21.21×10^{-15} and its uncertainty is 0.26×10^{-15}.
4. Spin exchange (type A). This frequency shift is produced by the collisions between the atoms. Due to the low atomic density, the effect is weak: -0.42×10^{-15} and its uncertainty is 0.10×10^{-15}.
5. Microwave leakage (type A). This effect arises if microwave radiation gets in the atom trajectory outside the interaction region. This effect contributes to both frequency inaccuracy and long-term instability since is not stable. Its mean value is null, the uncertainty is 0.20×10^{-15}.

6. ac Zeeman effect (type B). This Zeeman effect is due to the (weak) magnetic field produced by the heaters of the device. The fractional frequency bias is 0.05×10^{-15}, with an uncertainty of 0.05×10^{-15}.
7. Cavity pulling (type B). The mistuning of the resonant cavity where the atoms interact with the probing microwave radiation "pulls" the frequency of the transition, producing a frequency bias of 0.02×10^{-15}, with an uncertainty of 0.02×10^{-15}.
8. Rabi and Ramsey pullings (type B). These effects are due to the cesium atom ground state lines other than the clock lines [39, 38, 78]. Both effects are very weak in the case of [57]: frequency pulling and uncertainty of 1×10^{-19}.
9. Majorana transitions (type B). These transitions are caused by directional changes in the magnetic fields seen by the atoms along their trajectories [89, 123]. In the device of [57], the effect is weak, 0.02×10^{-15}, with an uncertainty of 0.02×10^{-15}.
10. Fluorescence light shift (type B). Before and after being probed by the microwave signal, the atoms are prepared and manipulated by various optical excitation. Any optical signal coming from these optical manipulations (for instance the fluorescence signal produced by the atoms themselves) modifies the atomic levels of the atoms being tested. In the case of an atomic fountain such as that of [57], the optical excitations are light-off when the atoms are probed, which reduces the effect to a very low and negligible level (1×10^{-20}).
11. Distributed cavity phase shift (type B). This effect results from the electrical losses in the cavity, which induce a traveling wave component. A value of 0.02×10^{-15} is assigned to this bias, with an identical uncertainty.
12. The second-order Doppler effect (type B). This relativistic effect is proportional to the ratio v_{a}^2/c^2, where v_{a} is the velocity of the atoms and c the speed of light. It is consequently very weak in the case of an atomic fountain: 0.02×10^{-15} in the case of [57].
13. dc Stark effect (type B). This effect is the splitting and shift of a spectral line into several components in the presence of an electric field [133]. This effect can be due to stray dc electric fields produced by thermoelectric currents, charged particles, macroscopic variations in the crystalline structure of the walls of the device. A careful design of the device eliminate this source of dc Stark effect. A worst-case evaluation in [57] gives a value of 0.02×10^{-15}.
14. Background gas collisions (type B). These collisions perturb the atomic levels and produce a shift of the hyperfine transition frequency. This shift has been studied for various gases [13, 6] and is proportional to the pressure. In the case of [57], the residual pressure is very low and the fractional shift is evaluated to be smaller than 0.001×10^{-15}.
15. The Bloch–Siegert effect (type B). This effect arises when the atoms interact with a non-rotating field, which is the case for the clock transition

of cesium [15]. The linear polarization of the microwave field results in the addition of two circular polarization counter-rotating fields, one of them producing the frequency bias. This frequency shift has been calculated [123] and gives a negligible bias: 0.0001×10^{-19}.

16. RF spectral purity (type B). The microwave power spectra introduces some frequency noise in the cavity, which can bias the frequency of the clock transition. For this reason, high quality frequency synthesis is required to obtain negligible uncertainty [56].The value reported in [57] is 3^{-18}.

In [57] the combination of these components gives a total uncertainty of 0.34×10^{-15}.

4.3 Typical Values

In this book, we are dealing with precise time and frequency applications. Therefore, we will only mention the typical parameters of the devices producing the best performances. Among them, the following two families must be considered:

- devices built and operated in a research laboratory and intended to have the best properties: the primary frequency standards; and
- devices built on a factory site, according to an industrial process, which can be operated by non-specialized staff. The accuracy and stability of these devices are not the ultimate ones, but their implementation and maintenance are easier and their reliability is better.

Both families of time and frequency standards are involved in the construction of the time scales (see Sect. 7.1) and some typical values of accuracy for both families are given below.

4.3.1 Primary Frequency Standards

Cesium atom fountains are the device which obtain the best values of uncertainty. These values are given in Table 4.1. More information can be found on the BIPM site [17].

The recently published uncertainty results of some thermal beam cesium primary frequency standards with optical pumping are given in Table 4.2.

Many other devices are under study, some of them being optical frequency standards. Their aim is to produce ultra-stable frequencies in the optical range by using a narrow optical transition as a clock transition.

One kind of device uses an optical transition of the calcium atom [97, 41]. The 10^{-15} relative uncertainty is under way [41].

Other devices use the optical transition of a single laser-cooled $^{199}Hg^{+}$ or $^{171}Yb^{+}$ ion confined in a radio frequency Paul trap [115, 99, 114]. Neutral ytterbium is also studied [61].

Table 4.1. Some values of accuracy obtained by cesium fountains

Institute	Device name	Accuracy	References
NIST	NIST-F1	0.35×10^{-15}	[57]
BNM-SYRTE	FO1	0.72×10^{-15}	[124]
PTB	CSF1	1×10^{-15}	[11]
IEN	CSF1	1.1×10^{-15}	[83, 10]
NPL	NPL-CsF1	1.2×10^{-15}	[10]
NMIJ	NMIJ-F1	4.2×10^{-15}	[74]

Table 4.2. Uncertainty of a few other primary frequency standards

Institute	Device	Device name	Accuracy	References
NIST	Cesium beam	NIST-7	5×10^{-15}	[112]
BNM-SYRTE	Cesium beam	JPO	6.3×10^{-15}	[90, 40]
NICT	Cesium beam	CRL-O1	7×10^{-15}	[55]
PRB	Cesium beam	CS1	9.5×10^{-15}	[23]

4.3.2 Commercial Devices

These devices are not designed in order to attain the best values of stability or accuracy. While keeping values of accuracy and stability compatible with most applications, they show a reliability that allows their use in a continuous way for long periods of time. For instance, the 5071A (Symmetricom) has demonstrated an average mean time between failures (MTBF) greater than 160 000 hours (which corresponds to more than 18 years) since its introduction in 1992 [113]. This property is essential to maintain any time scale, and is used to build TAI (see Sect. 7.1.1).

The typical accuracy of the commercially available cesium beams frequency standards is from a few parts in 10^{13} to 1×10^{-12} [113, 2].

5 Reproducibility

The reproducibility of a device or of a measurement is its ability to produce, repeatedly and without adjustments, the same value or result, given the same input conditions and operating in the same environment.

In the case of an oscillator, the reproducibility is characterized by the dispersion of the values of the frequency delivered by the device each time it is turned on.

This property is closely related to the long term stability, the sensitivity to the environment and the accuracy.

It is an important parameter for commercial devices that may be used by non-specialists.

For instance, the reproducibility of the commercial cesium atomic frequency standard 5071A (Symmetricom) given in the data sheet [113] is $\pm 1 \times 10^{-13}$.

This means that, what the environment conditions in the limits of the specifications may be, the frequency error is smaller than this value.

6 Stability

6.1 Definition

While accuracy (uncertainty) indicates whether the value is right or wrong, stability indicates whether the value stays the same over a given time interval and *determines how well it can produce the same frequency over a given time interval* [96].

For instance, if the frequency of a 10 MHz quartz crystal oscillator varies between 9.999995 and 9.999996 MHz over a given time interval, we could say that its stability is 1 Hz on that time interval. It is most often expressed as a fractional dimensionless value by dividing the frequency fluctuations value by the nominal frequency of the oscillator.

In this example, the fractional stability is $\delta f/f_n = 1 \times 10^-7$.

Notice that a low value corresponds to a good stability. In fact, the value of the frequency of an oscillator varies with time for the following reasons.

1. Random fluctuations. The instantaneous frequency deviation due to these fluctuations is denoted by $\Delta f(t)$. It is a random function of time, which is assumed to be second-order stationary, the mean value and the covariance of the fluctuations are time-invariant.
   These random frequency fluctuations can be written as
$$\Delta f(t) = \delta f_0 + \delta f(t) \,, \tag{6.1}$$
$$\delta f_0 = E\{\Delta f(t)\} \,, \tag{6.2}$$
$$E\{\delta f(t)\} = 0 \,, \tag{6.3}$$
   where $\Delta f(t)$ is the instantaneous value, at time t, of the frequency deviation and δf_0 is the mean value of the frequency deviation.
   The constant part of $\Delta f(t)$ contributes to the frequency inaccuracy and only the fluctuating part is considered in frequency stability studies.
2. Deterministic variations of the frequency. These may be due to known processes, such as the modification of the environment of the device, or to uncontrolled processes, such as aging, which can produce a low frequency drift. This drift is assumed to be linear with time t,
$$\delta f_d(t) = k(t - t_0) \,. \tag{6.4}$$

In this expression, t_0 is the time where the effect of the drift cancels. Due to frequency drift, the overall frequency deviation is not stationary, which leads to some difficulties, since the Fourier transform of this overall frequency deviation cannot be defined.

The instantaneous frequency of the oscillator under test is, therefore,

$$f(t) = f_0(t) + \delta f_0 + \delta f(t) + k(t - t_0) \, . \tag{6.5}$$

Characterization of the frequency stability should give pertinent information about the random process $\delta f(t)$ and the continuous frequency drift $k(t - t_0)$.

The following two different families of tools are involved in such studies:

1. Measurement in the frequency domain (spectrum analyzer).
   The frequency fluctuations are converted to a voltage fluctuation, which is analyzed in order to determine its properties.
   Such precise frequency measurements imply that
   - the frequency of the oscillator under test is transposed toward lower frequencies without loss of its stability, which means that the local oscillator must be at least as good as the oscillator under test, and
   - the voltage fluctuations must be studied very close to the null frequency.

   This technique will not be studied herein. It is best suited for very short term stability (high values of the Fourier frequency of the frequency noise spectrum), which are not in the scope of the book.
2. Measurement in the time domain. This technique allows one to obtain information on the medium and long-term stability (very low Fourier frequencies of the frequency noise spectrum).
   The frequency of the oscillator is measured over time intervals of duration τ by counting the number of periods during these time intervals via a frequency meter or period counter. The sequence of discrete data resulting from these measurements is analyzed.
   Such precise frequency measurements imply that
   - the reference clock of the frequency or period meter must be at least as good as the oscillator under test in order to perform precise frequency measurements;
   - very low frequency fluctuations are to be measured, which implies long measurement times; and
   - the data analyzed result from processing the instantaneous frequency of the oscillators as described in Sect. 6.2.1. This process filters the instantaneous frequency fluctuations and samples the result. There may consequently be a loss of information.

A full characterization of the stability of an oscillator should give the precise power spectra of the random frequency fluctuations and the drifts that affect its accuracy.

6.2 Measurements in the Time Domain

The frequency of the oscillator is measured over time intervals of duration τ by counting the number of periods during these time intervals. The result is a sequence of numerical values, each one being the mean value of the frequency of the oscillator (including the random fluctuations) during the corresponding time interval. This discrete-time random process is then studied.

6.2.1 The Measurement Process

The quantity to be characterized is the continuous random process $\delta f(t)$. The raw data given by the process of counting the periods of the down converted signal are the results given by the frequency meter at the end of each measurement of duration τ. These measurements give the mean value of the frequency during this duration, since it is the ratio of the number of periods of the signal during the time τ and of this duration τ. The consequences are as follows:

1. The continuous-time random process $\delta f(t)$ cannot be reached directly. We call $\delta f_\tau(t_n)$ the result of a frequency measurement obtained at time t_n. It is the mean value of the instantaneous frequency over the time interval $[t_n - \tau, t_n]$ (τ is the integration time)

$$f_\tau(t_n) = \frac{1}{\tau}\int_{t_n-\tau}^{t_n} f(t)\mathrm{d}t \,. \tag{6.6}$$

2. The continuous-time process $\delta f_\tau(t)$ cannot be reached directly either. Instead, the successive results of the frequency measurement over the time duration τ are *samples* of that continuous time function $f_\tau(t)$ at the sampling times $t_n = n \times T$, where T is the sampling period and $1/T = \nu_\mathrm{s}$ is the sampling rate or sampling frequency

$$\delta f_\tau(n) = \delta f_\tau(nT) \,. \tag{6.7}$$

The sampling period and the integration time are related by

$$T \geq \tau \,. \tag{6.8}$$

The different steps of the process are described below and their consequences are studied.

The Frequency Measurement Filters the Frequency Fluctuations

The continuous-time process $f_\tau(t)$ is deduced from the frequency fluctuations $\delta f(t)$ by a filter defined by (6.6).

The impulse response $h_\mathrm{m}(t)$ of this filter is easily deduced from (6.6)[1]

$$h_\mathrm{m}(t) = \begin{cases} 1/\tau \; 0 < t < \tau \\ 0 \quad t < 0, \; t > \tau \end{cases} . \tag{6.9}$$

The transfer function $H_\mathrm{m}(p)$ of this filter is the Laplace transform of this impulse response

$$H_\mathrm{m}(p) = \frac{1}{\tau p}\left(1 - \mathrm{e}^{\tau p}\right) . \tag{6.10}$$

Its frequency response $G_\mathrm{m}(\nu)$ is

$$G_\mathrm{m}(\nu) = \mathrm{e}^{-j2\pi\nu\tau/2} \times \frac{\sin 2\pi\nu\tau/2}{2\pi\nu\tau/2} , \tag{6.11}$$

where ν is the Fourier frequency.

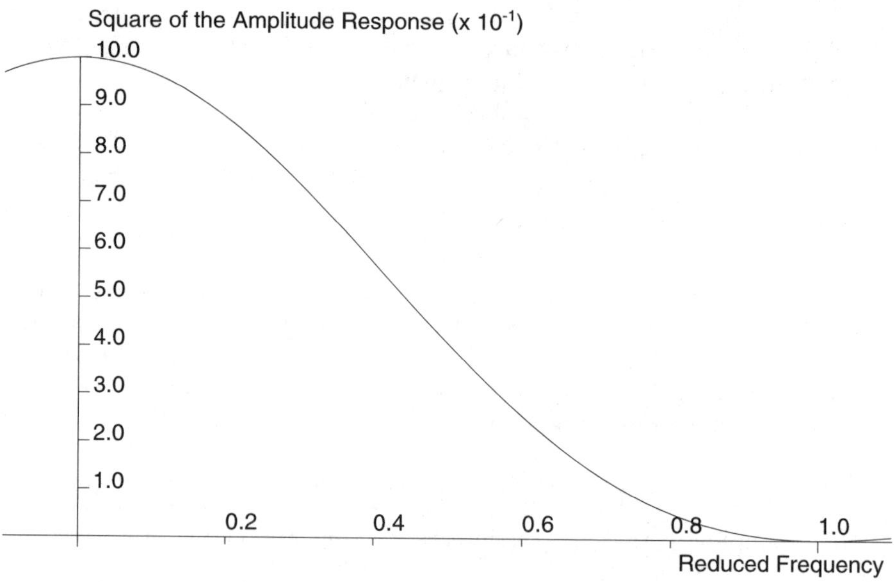

Fig. 6.1. Squared amplitude response of the filter versus the reduced frequency

[1] Remember that the output $y(t)$ of a filter whose impulse response is $h(t)$ is

$$y(t) = \int_{-\infty}^{\infty} x(t - \tau) \times h(\tau)\mathrm{d}\tau ,$$

where $x(t)$ is the input of the filter.

Its amplitude response is

$$|G_{\mathrm{m}}(\nu)| = \left| \frac{\sin \pi\nu\tau}{\pi\nu\tau} \right| . \tag{6.12}$$

This is a low-pass filter with a unit gain at the null frequency and a first zero transmission at the Fourier frequency,

$$\nu_0 = \frac{1}{\tau} . \tag{6.13}$$

As a function of the reduced Fourier frequency $\nu_\tau = \nu \times \tau$, the amplitude response $G_{\mathrm{m}}^{\tau}(\nu_\tau)$ is

$$G_{\mathrm{m}}^{\tau}(\nu_\tau) = \left| \frac{\sin \pi\nu_\tau}{\pi\nu_\tau} \right| . \tag{6.14}$$

Figure 6.1 shows the squared amplitude response $|G_{\mathrm{m}}^{\tau}(\nu_\tau)|^2$ of this filter versus the reduced frequency $\nu_\tau = \nu \times \tau$.

Sampling Rate Versus Integration Time

Due to the aliasing effect (see Sect. 6.2.1), it is important to compare the sampling rate and the band of the low-pass filter. The sampling period T cannot be smaller than the duration τ of an individual measurement, since no data is given before the end of the time τ.

Three situations can consequently be considered.

1. No dead time in the frequency measurement.
   In this case, a new frequency measurement is started at time nT when a result is given, and the next value will be given at time $(n+1)T = nT+\tau$. We then have in this case

$$T = \tau , \tag{6.15}$$

$$\nu_{\mathrm{s}} = \frac{1}{\tau} . \tag{6.16}$$

   The frequency response of the filter produced by the measurement can be written as a function $\mathcal{G}_m$ of the reduced frequency $\nu_r = \nu/\nu_s$

$$\mathcal{G}_{\mathrm{m}}(\nu_{\mathrm{r}}) = \mathrm{e}^{-j2\pi\nu_{\mathrm{r}}\tau/(2T)} \times \frac{\sin\{2\pi\nu_{\mathrm{r}}\tau/(2T)\}}{2\pi\nu_{\mathrm{r}}\tau/(2T)} \tag{6.17}$$

$$= \mathrm{e}^{-j\pi\nu_{\mathrm{r}}} \times \frac{\sin \pi\nu_{\mathrm{r}}}{\pi\nu_{\mathrm{r}}} . \tag{6.18}$$

   Figure 6.2 shows the square of the amplitude response as a function of the reduced frequency ν_{r}.
   The process of measuring successive values of the mean frequency during time intervals τ with no dead time consequently leads to filtering of the spectrum of the frequency fluctuations with a low-pass filter whose first zero transmission occurs at the sampling frequency $\nu_{\mathrm{s}} = 1/T = 1/\tau$.

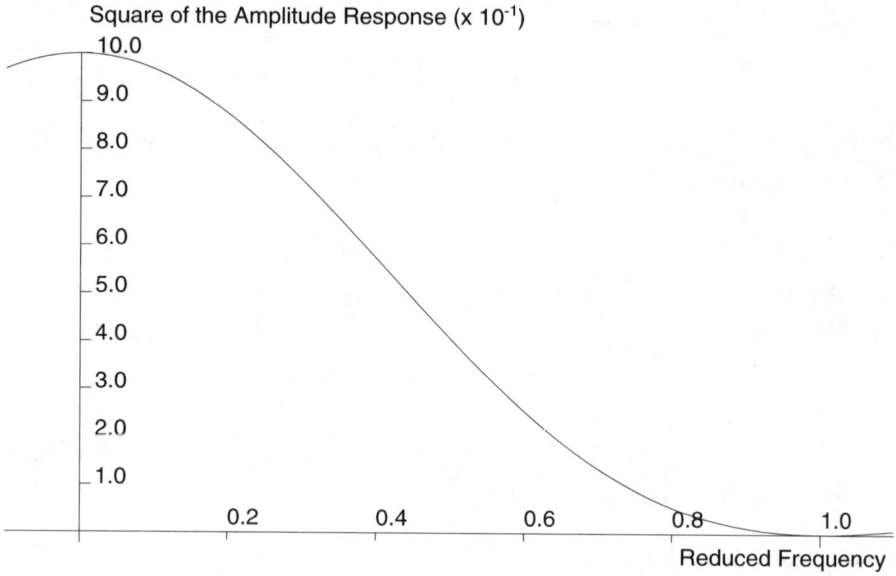

Fig. 6.2. Squared amplitude response of the filter associated to the frequency measurement with no dead time

2. Dead time of duration τ' in the frequency measurement.
   In this case, the next frequency measurement begins after the dead time τ' and the value will be obtained after the time $T = \tau + \tau'$ (see Fig. 6.3). We have

$$T > \tau \, , \tag{6.19}$$

$$\nu_\mathrm{s} < \frac{1}{\tau} \, . \tag{6.20}$$

The function $\mathcal{G}_\mathrm{m}$ of the reduced frequency $\nu_\mathrm{r} = \nu/\nu_\mathrm{s}$ is

$$\mathcal{G}_\mathrm{m}(\nu_\mathrm{r}) = \mathrm{e}^{-j\pi\nu_\mathrm{r}/(1+k)} \times \frac{\sin\{\pi\nu_\mathrm{r}/(1+k)\}}{\pi\nu_\mathrm{r}/(1+k)} \, . \tag{6.21}$$

k is the ratio of the dead time to the integration time.

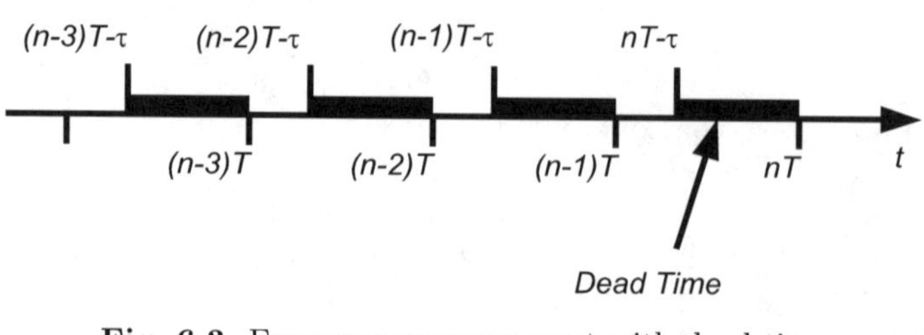

Fig. 6.3. Frequency measurement with dead time

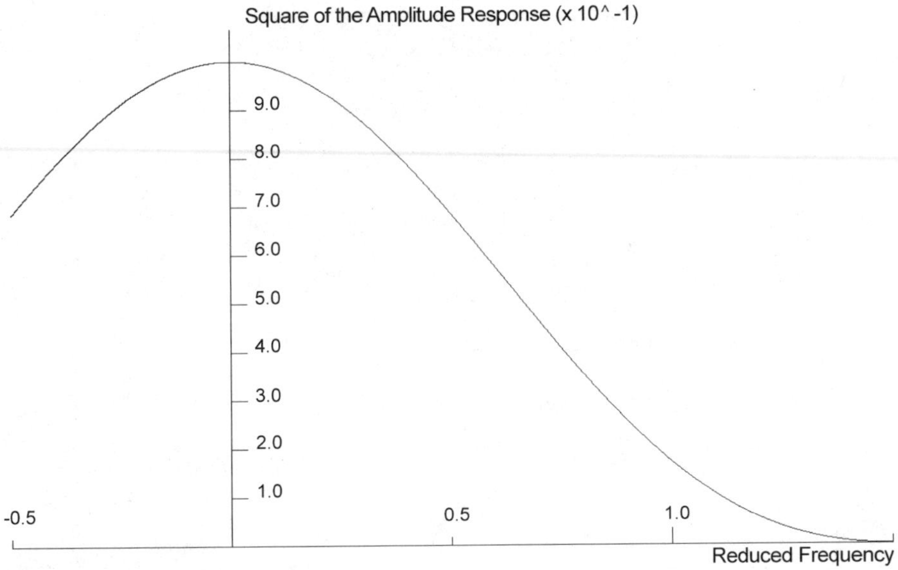

Fig. 6.4. Squared amplitude response of the filter associated to the frequency measurement with a dead time

Figure 6.4 shows the squared amplitude response as a function of the reduced frequency ν_{r} in the case $\tau' = \tau/2$.
The process of measuring successive values of the mean frequency during time intervals τ with a dead time $\tau\prime$ then leads to filtering of the spectrum of the frequency fluctuations with a low-pass filter whose first zero transmission occurs beyond the sampling frequency $\nu_{\mathrm{s}} = 1/T = 1/(\tau + \tau\prime)$.

3. No dead time and association of successive results.
In this case, M successive samples of duration τ_0 are used to calculate the mean value $f_{\mathrm{M}\tau_0}(t)$ of the instantaneous frequency $f(t)$ over a duration $M\tau_0 = \tau$

$$f_\tau(t_0) = \frac{1}{\tau}\int_{t_0-\tau}^{t_0} f(t)\mathrm{d}t \tag{6.22}$$

$$= \frac{1}{M\tau_0}\sum_{m=0}^{M-1}\left\{\int_{t_0-(m+1)\tau/M}^{t_0-m\tau/M} f(t)\mathrm{d}t\right\} \tag{6.23}$$

$$= \frac{1}{M\tau_0}\sum_{m=0}^{M-1} f_{\tau/\mathrm{M}}(t_0 - m\tau/M)\,. \tag{6.24}$$

One mean frequency measurement is the concatenation of M elementary measurements. Its duration is $\tau = M\tau_0$, but samples are taken at the end of every elementary measurement, with a sampling rate of $\nu_{\mathrm{s}} = 1/T = 1/\tau_0$.

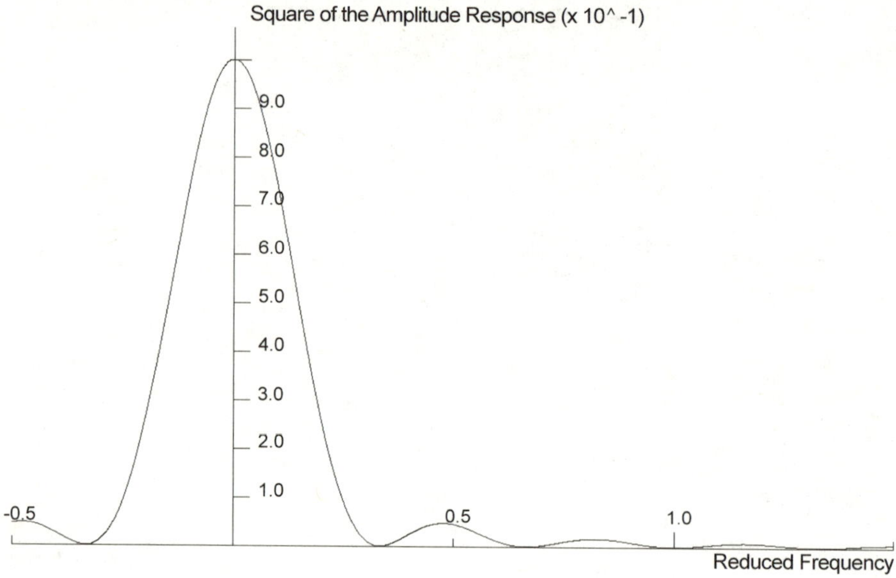

Fig. 6.5. Squared amplitude response of the filter associated to the frequency measurement with no dead time and association of three successive results

We then have

$$T = \frac{\tau}{M} , \tag{6.25}$$

$$\nu_s = M \times \frac{1}{\tau} . \tag{6.26}$$

The function $\mathcal{G}_m$ of $\nu_r = \nu/\nu_s$ is in this case

$$\mathcal{G}_m(\nu_r) = e^{-j\pi N \nu_r} \times \frac{\sin\{\pi N \nu_r\}}{\pi N \nu_r} . \tag{6.27}$$

Figure 6.5 shows the squared amplitude response as a function of the reduced frequency ν_r in the case $N = 3$.
The process of measuring successive values of the mean frequency during time intervals τ_0 with no dead time and associating N successive results to construct one sample of the sequence consequently leads to filtering of the spectrum of the frequency fluctuations with a low-pass filter whose first zero transmission occurs at the sampling frequency divided by N.

Frequency Aliasing

It is well known (see for instance [131]) that sampling a continuous time function $x_c(t)$ whose Fourier transform is $X_c(\nu)$ with a sampling time $T = 1/\nu_s$

leads to a sequence $x(n) = x_c(nT)$ whose Fourier transform $X(\nu)$ is deduced from $X_c(\nu)$ by (the Nyquist–Shannon sampling theorem):

$$X(\nu) = \frac{1}{T} \sum_{k=-\infty}^{\infty} X_c\left(\nu + k\nu_s\right) . \tag{6.28}$$

In such a situation, the Fourier transform of the continuous signal, and consequently the continuous signal itself, cannot be computed from the Fourier transform of the discrete signal: there has been a loss of information.

Consequently, the Fourier transform $X(\nu)$ of the sampled signal in the Fourier frequency domain $[0 - \nu_s/2]$ is different from the Fourier transform $X_c(\nu)$ of the continuous signal in this domain if $X_c(\nu)$ spreads beyond the half of the sampling frequency $\nu_s/2$ (the Nyquist frequency), as shown in Fig. 6.6. This is the aliasing phenomenon.

In most usual cases where the spectral density of the continuous signal is a quantity of interest, either

- the sampling frequency ν_s is chosen large enough so that the spectrum of interest does not extend beyond the Nyquist frequency, or
- the part of that spectrum that extends beyond the Nyquist frequency is suppressed by a low-pass filter *before* sampling.

The filter effect produced by the frequency measurement does not limit the frequency domain of the Fourier transform of the frequency noise, unless we

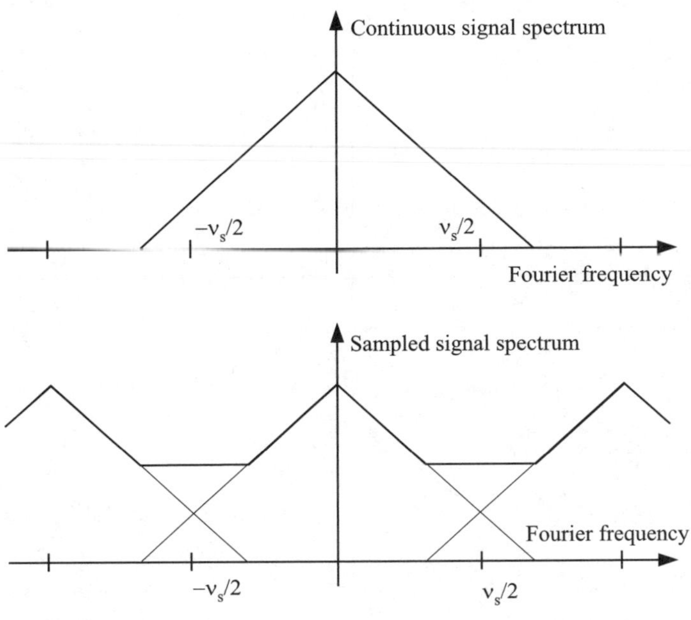

Fig. 6.6. Spectrum of the discrete-time process compared to that of the continuous-time process

neglect the side lobes of $G(\nu)$ (or $\mathcal{G}_{\mathrm{m}}(\nu_{\mathrm{r}})$). It is consequently not possible to deduce the Fourier transform (or the spectral density) from the sequence made from the results of the successive frequency measurement.

Only in the case where N successive measurements are associated, the main lobe of $G(\nu)$ is inside the Nyquist frequencies $\pm\nu_{\mathrm{s}}/2$, since $N \geq 2$.

In the general case, the part of the Fourier transform of $f_{\tau}(t)$ beyond the Nyquist frequency is folded back in the band $0 \leq \nu \leq \nu_{\mathrm{s}}/2$ of the Fourier transform of $f_{\tau}(n)$ (see Fig. 6.7).

In the simple case where the Fourier transform of the continuous-time process $f_{\tau}(t)$ does not extends beyond twice the Nyquist frequency, we have for every value of ν such as $0 \leq \nu \leq \nu_{\mathrm{s}}/2$,

$$F_{\tau}(\nu) = F_{\tau,c}(\nu) + F_{\tau,c}(\nu_{\mathrm{s}} - \nu) \,. \tag{6.29}$$

The quantity of interest is the spectral density of the discrete signal

$$\begin{aligned}
P_{F,\tau}(\nu) &= F_{\tau}(\nu) \times F_{\tau}^{*}(\nu) & (6.30)\\
&= |F_{\tau,c}(\nu)|^2 + |F_{\tau,c}(\nu_{\mathrm{s}} - \nu)|^2 & (6.31)\\
&\quad + |F_{\tau,c}(\nu)|\,|F_{\tau,c}(\nu_{\mathrm{s}} - \nu)| \exp\left[j\left(\phi(\nu) - \phi(\nu_{\mathrm{s}} - \nu)\right)\right] & (6.32)\\
&\quad + |F_{\tau,c}(\nu)|\,|F_{\tau,c}(\nu_{\mathrm{s}} - \nu)| \exp\left[j\left(-\phi(\nu) + \phi(\nu_{\mathrm{s}} - \nu)\right)\right] & (6.33)\\
&= |F_{\tau,c}(\nu)|^2 + |F_{\tau,c}(\nu_{\mathrm{s}} - \nu)|^2 & (6.34)\\
&\quad + 2\,|F_{\tau,c}(\nu)|\,|F_{\tau,c}(\nu_{\mathrm{s}} - \nu)| \cos(\delta\phi) \,, & (6.35)
\end{aligned}$$

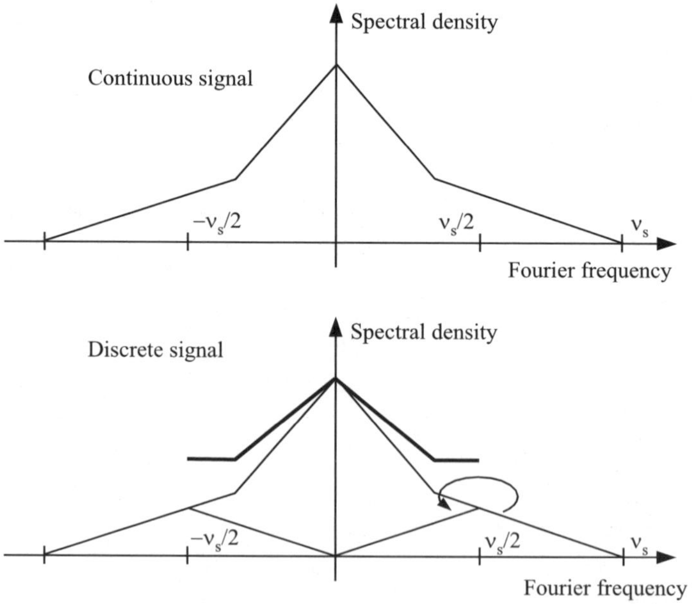

Fig. 6.7. Aliasing when frequency is measured without a dead time

where $\delta\phi$ is the phase difference between the Fourier transform values for the Fourier frequencies ν and $\nu_s - \nu$. If we assume that the phase of the Fourier transform of the time-continuous process at a frequency $0 \leq \nu \leq< \nu_s/2$ and the phase at the frequency $\nu\prime = \nu_s - \nu$ are not correlated, the value of $\cos(\delta\phi)$ can be averaged to zero and we can approximate

$$P_{F,\tau}(\nu) = |F_{\tau,c}(\nu)|^2 + |F_{\tau,c}(\nu_s - \nu)|^2 \ . \tag{6.36}$$

The conclusion is that the power of the discrete-time process calculated for the Fourier frequencies inside the Nyquist value equals the total power of the continuous-time process.

This result can easily be generalized for any value of the maximal Fourier frequency of the continuous-time process.

The main information given by frequency measurement is consequently the value of the power of the frequency noise filtered by the low-pass filter equivalent to the measurement process.

6.2.2 Power and Spectral Density

The power of the random part $\delta f(t)$ of the frequency of the oscillator under test is a characteristic of the stability. This power is given by the variance of this random process. Some other statistical parameters may be useful and are described below.

Statistical Properties of Random Processes

Ergodic and Stationary Random Processes

Here we are concerned with an ergodic [127] and second-order stationary [134] random process $\{X_t\}$. Its properties are the following:

1. The mean value $\mu(t)$ of the process is time-invariant. This defines first-order stationarity

$$\mu(t) = E\{X_t\} = \mu \ . \tag{6.37}$$

   In the following, we remove the constant mean value of the process $\mu = 0$.
2. The covariance $\gamma(t2, t1)$ of the process is time-invariant and is only a function of the difference $t2 - t1$. This defines second-order stationarity

$$\gamma(t2, t1) = E\{(X_{t2})(X_{t1})^*\} = \gamma(t2 - t1) \ . \tag{6.38}$$

3. The variance $\sigma^2(t)$ of the process, which is its covariance for $t2 - t1 = 0$, is time-invariant

$$\sigma^2(t) = E\{|X_t|^2\} = \sigma^2 \ . \tag{6.39}$$

   These definitions hold for both discrete-time and continuous-time processes.

a) For a continuous-time process $x(t)$

$$E\{x(t)\} = \mu = 0\ , \tag{6.40}$$

$$E\{[x(t2)][x(t1)]^*\} = \gamma(t2 - t1)\ , \tag{6.41}$$

$$E\{|x(t)|^2\} = \sigma^2\ . \tag{6.42}$$

b) For a discrete-time process $x(n)$

$$E\{x(n)\} = \mu = 0\ , \tag{6.43}$$

$$E\{[x(n2)][x(n1)]^*\} = \gamma(n2 - n1)\ , \tag{6.44}$$

$$E\{|x(n)|^2\} = \sigma^2\ . \tag{6.45}$$

4. The time average and the average over the statistical ensemble give the same result: this defines ergodicity.
   This means that, for instance,

$$E\{X_t\} = \lim_{\Delta T\to\infty}\left\{\frac{1}{2\Delta T}\int_{-\Delta T}^{\Delta T} X_t \mathrm{d}t\right\} \tag{6.46}$$

   if X_t is a continuous-time process, and

$$E\{X_t\} = \lim_{N\to\infty}\left\{\frac{1}{2N+1}\sum_{-N}^{N} X_t\right\} \tag{6.47}$$

   if it is a discrete-time process.

Spectral Density and Variance

The spectral density $p(\nu)$ of a stationary random process X_t is defined as the Fourier transform of its covariance.

1. For a continuous-time process

$$p(\nu) = \int_{-\infty}^{\infty} \gamma(t)\exp(-j2\pi\nu t)\mathrm{d}t\ , \tag{6.48}$$

$$\gamma(t) = \int_{-\infty}^{\infty} p(\nu)\exp(j2\pi\nu t)\mathrm{d}\nu\ . \tag{6.49}$$

2. For a discrete-time process

$$p(\nu) = T\sum_{n=-\infty}^{\infty} \gamma(n)\exp(-j2\pi\nu nT)\ , \tag{6.50}$$

$$\gamma(n) = \int_{-1/2T}^{1/2T} p(\nu)\exp(j2\pi\nu nT)\mathrm{d}\nu\ , \tag{6.51}$$

   where T is the sampling time.

If the dimension of the quantity X_t is dim (for instance, in the case of frequency fluctuations, the dimension of the random process is Hz), the dimension of the covariance is dim^2 (Hz^2 in the case of frequency fluctuations).

The dimension of the spectral density is therefore, $\mathrm{dim}^2 \times$ s, or $\mathrm{dim}^2 \times \mathrm{Hz}^{-1}$. In the case of frequency fluctuations, the dimension of the spectral density is consequently Hz.

The process variance σ^2, which is its mean power, can be computed from the spectral density.

1. For a continuous-time process

$$\sigma^2 = E\{x(t)^2\} \tag{6.52}$$
$$= \gamma(0) \tag{6.53}$$
$$= \int_{-\infty}^{\infty} p(\nu)\mathrm{d}\nu \,. \tag{6.54}$$

2. For a discrete-time process

$$\sigma^2 = E\{x(n)^2\} \tag{6.55}$$
$$= \gamma(0) \tag{6.56}$$
$$= \int_{-1/2T}^{1/2T} p(\nu)\mathrm{d}\nu \,. \tag{6.57}$$

The spectral density indicates how the power of the process is distributed among its different Fourier frequencies. It can also be estimated from the Fourier transform $X(\nu)$ of the signal.

Assuming the ergodic hypothesis, $p(\nu)$ can be written as follows:

1. For a continuous-time process

$$p(\nu) = \lim_{\Delta T \to \infty} \frac{1}{2\Delta T} \left| \int_{-\Delta T}^{\Delta T} x(t) \exp(-2j\pi\nu t)\mathrm{d}t \right|^2 \,. \tag{6.58}$$

2. For a discrete-time process

$$p(\nu) = T \lim_{N \to \infty} \frac{1}{2N+1} \left| \sum_{n=-N}^{N} x(n) \exp(-2j\pi\nu nT) \right|^2 \,. \tag{6.59}$$

The quantity

$$\int_{-\Delta T}^{\Delta T} x(t) \exp(-2j\pi\nu t)\mathrm{d}t \,,$$

which appears on the right-hand side of (6.58) is the Fourier transform of the continuous-time process $x(t)$ in the rectangular window $[-\Delta T, +\Delta T]$.

The quantity

$$\sum_{n=-N}^{N} x(n) \exp(-2j\pi\nu nT)\mathrm{d}t ,$$

which appears on the right-hand side of (6.59) is the Fourier transform of the discrete-time process $x(n)$ in the rectangular window $[-N, +N]$.

In fact, any measurement will be limited to a finite duration $2\Delta T$ (for a continuous-time process) or to a finite number $2N+1$ of samples (for a discrete-time process) and the power spectral density will be estimated by the quantity p_{per}, which is often called a periodogram.

1. For a continuous-time process

$$p_{\mathrm{per}}(\nu) = \frac{1}{2\Delta T}\left|\int_{-\Delta T}^{\Delta T} x(t) \exp(-2j\pi\nu t)\mathrm{d}t\right|^2 . \tag{6.60}$$

2. For a discrete-time process

$$p_{\mathrm{per}}(\nu) = T\frac{1}{2N+1}\left|\sum_{n=-N}^{N} x(n) \exp(-2j\pi\nu nT)\right|^2 . \tag{6.61}$$

The Discrete-time Process Versus the Continuous-time Process

It is useful to relate the spectral density of a discrete-time process to the spectral density of the continuous-time process from which it is deduced by sampling. Let $x(t)$ a continuous-time process, ergodic and second-order-stationary, and $x(n)$ the discrete-time process defined by sampling $x(t)$ with a sampling rate $\nu_{\mathrm{s}} = 1/T$. Then

$$x(n) = x(nT) . \tag{6.62}$$

The covariance and the spectral density of $x(t)$ are, respectively, $\gamma(t)$ and $p(\nu)$, and the covariance and the spectral density of $x(n)$ are, respectively, $\Gamma(n)$ and $P(\nu)$.

The covariances and spectral densities are related by

$$p(\nu) = \int_{-\infty}^{\infty} \gamma(\tau) \exp(-j2\pi\nu\tau)\mathrm{d}\tau , \tag{6.63}$$

$$\gamma(\tau) = \int_{-\infty}^{\infty} p(\nu) \exp(j2\pi\nu\tau)\mathrm{d}\nu \tag{6.64}$$

and

$$P(\nu) = T\sum_{-\infty}^{\infty} \Gamma(n) \exp(-j2\pi\nu nT) , \tag{6.65}$$

$$\Gamma(n) = \int_{-1/2T}^{1/2T} P(\nu) \exp(j2\pi\nu nT)\mathrm{d}\nu . \tag{6.66}$$

The covariance is defined by

$$\gamma(\tau) = E\{x(t)x^*(t-\tau)\}\ , \tag{6.67}$$

$$\Gamma(n) = E\{x(mT)x^*(mT-nT)\} \tag{6.68}$$

$$= \gamma[nT]\ . \tag{6.69}$$

Consequently,

$$\Gamma(n) = \int_{-\infty}^{\infty} p(\nu)\exp(j2\pi\nu nT)\mathrm{d}\nu \tag{6.70}$$

$$= \sum_{k=-\infty}^{\infty} \int_{(k-1/2)\nu_\mathrm{s}}^{(k+1/2)\nu_\mathrm{s}} p(\nu)\exp(j2\pi\nu nT)\mathrm{d}\nu\ . \tag{6.71}$$

Putting $f = \nu - k/T = \nu - k\nu_\mathrm{s}$, we have

$$\Gamma(n) = \sum_{k=-\infty}^{\infty} \int_{-\nu_\mathrm{s}/2}^{\nu_\mathrm{s}/2} p(f+k\nu_\mathrm{s})\exp[j2\pi(f+k/T)nT]\mathrm{d}\nu \tag{6.72}$$

$$= \sum_{k=-\infty}^{\infty} \int_{-\nu_\mathrm{s}/2}^{\nu_\mathrm{s}/2} p(f+k\nu_\mathrm{s})\exp(j2\pi fnT)\mathrm{d}\nu \tag{6.73}$$

$$= \int_{-\nu_\mathrm{s}/2}^{\nu_\mathrm{s}/2} \sum_{k=-\infty}^{\infty} p(f+k\nu_\mathrm{s})\exp(j2\pi fnT)\mathrm{d}\nu\ . \tag{6.74}$$

Comparing (6.66) and (6.74) gives

$$P(\nu) = \sum_{k=-\infty}^{\infty} p(\nu + k\nu_\mathrm{s})\ . \tag{6.75}$$

Consequently, we have

$$\int_{-\nu_\mathrm{s}/2}^{\nu_\mathrm{s}/2} P(\nu)\mathrm{d}\nu = \int_{-\nu_\mathrm{s}/2}^{\nu_\mathrm{s}/2} \sum_{k=-\infty}^{\infty} p(\nu+k\nu_\mathrm{s})\mathrm{d}\nu \tag{6.76}$$

$$= \sum_{k=-\infty}^{\infty} \int_{-\nu_\mathrm{s}/2}^{\nu_\mathrm{s}/2} p(\nu+k\nu_\mathrm{s})\mathrm{d}n\nu\ . \tag{6.77}$$

Putting $f = \nu + k\nu_\mathrm{s}$ gives

$$\int_{-\nu_\mathrm{s}/2}^{\nu_\mathrm{s}/2} P(\nu)\mathrm{d}\nu = \sum_{k=-\infty}^{\infty} \int_{(k-1/2)\nu_\mathrm{s}}^{(k+1/2)\nu_\mathrm{s}} p(f)\mathrm{d}f \tag{6.78}$$

$$= \int_{-\infty}^{\infty} p(f)\mathrm{d}f\ . \tag{6.79}$$

Consequently,

- the variance of the discrete-time process equals that of the continuous-time process, and
- the range of Fourier frequencies $[-\nu_s/2, \nu_s/2]$ of the discrete-time process contains the power of the range $[-\infty, \infty]$ of the continuous-time process.

Filtering

Filtering a random process X_t whose spectral density is $p_x(\nu)$ with a filter whose frequency response is $H(\nu)$ results in a random process Y_t whose spectral density is $p_y(\nu)\,|H(\nu)|^2$.

Statistical Properties of Frequency and Phase Fluctuations

We assume that the frequency fluctuation $\delta f(t) = f(t) - E\{f(t)\}$ is a continuous-time, stationary and ergodic random process, as well as the result of frequency measurement $\delta f_\tau(n) = f_\tau(n) - E\{f_\tau(n)\}$ is a discrete-time, stationary and ergodic random process.

The properties described above can consequently be applied to these two processes.

The Power of Frequency and Phase Fluctuations

The power of the random frequency fluctuations $\delta f(t) = f(t) - E\{f(t)\}$, defined as

$$P_{\delta f} = E\{(\delta f(t))^2\} \tag{6.80}$$

characterizes the strength of the frequency instability. We will consequently try to use this parameter to characterize the stability of the oscillator under test.

Assuming the ergodic hypothesis, we can write

$$P_{\delta f} = E\{(\delta f(t))^2\} \tag{6.81}$$

$$= \sigma_{\delta f}^2 \tag{6.82}$$

$$= \lim_{\Delta t \to \infty} \frac{1}{2\Delta t} \int_{-\Delta t}^{\Delta t} \delta f(t)^2 \mathrm{d}t \,. \tag{6.83}$$

Equivalent relations hold for the discrete-time process $\delta f_\tau(n) = f_\tau(n) - E\{f_\tau(n)\}$,

$$P_{\delta f,\tau} = E\{(\delta f_\tau(n))^2\} \tag{6.84}$$

$$= \sigma_{\delta f,\tau}^2 \tag{6.85}$$

$$= \lim_{N \to \infty} \frac{1}{2N+1} \sum_{-N}^{N} \delta f_\tau(n)^2 \,. \tag{6.86}$$

In fact, the quantity of interest is the dimensionless reduced (or fractional) frequency fluctuation

$$y(n) = \delta f_\tau(n)/f_0 \,, \tag{6.87}$$

where f_0 is the nominal frequency of the oscillator under test.

The reduced frequency fluctuations are very small for an ultra-stable oscillator and the variance, frequently denoted as σ_y^2, or any other parameter that characterizes the random process, is an extremely small number; the smaller it is, the smaller the frequency fluctuations and the better the stability of the oscillator.

The random processes associated to this quantity are themselves dimensionless, as well as their variance.

The dimension of the spectral density of the reduced frequency fluctuation is Hz^{-1}.

Another quantity frequently involved in the characterization of the frequency stability is the phase deviation from the nominal value $\phi_0(t) = \phi(0) + 2\pi f_0 t$ (phase fluctuations, phase noise).

The phase $\phi(t)$ of the signal delivered by the oscillator at time t is

$$\phi(t) = \phi(0) + 2\pi \int_0^t f(t)\mathrm{d}t \tag{6.88}$$

$$= \phi(0) + 2\pi f_0 t + 2\pi \int_0^t \delta f(t)\mathrm{d}t \tag{6.89}$$

$$= \phi_0(t) + \delta\phi(t) \,, \tag{6.90}$$

where $\phi_0(t) = \phi(0) + 2\pi f_0 t$ is the nominal phase corresponding to the nominal frequency f_0 of the oscillator. $\delta\phi(t)$ is the phase deviation at time t.

The phase deviation and the frequency deviation are related by

$$\delta f(t) = \frac{\mathrm{d}\delta\phi(t)}{\mathrm{d}t} \,, \tag{6.91}$$

$$\delta\phi(t) = 2\pi \int_0^t \delta f(t)\mathrm{d}t \,. \tag{6.92}$$

As for the frequency, the reduced phase deviation is the quantity of interest. It is defined by

$$x(t) = \delta\phi(t)/f_0 \,. \tag{6.93}$$

The dimension of this quantity is Hz^{-1}, the dimension of its variance is Hz^{-2}, and the dimension of its spectral density is Hz^{-3}.

Spectral Density of Frequency and Phase Fluctuations

A precise characterization of the stability is given by the spectral density of $\delta f(t) = f(t) - E\{f(t)\}$ or $\delta f_\tau(n) = f_\tau(n) - E\{f_\tau(n)\}$.

This indicates how the power of the random process $\delta f(t)$ or $\delta f_\tau(n)$ is distributed among its Fourier frequencies.

According to (6.48) and (6.50), the spectral density of the reduced frequency fluctuations is

$$p_y(\nu) = \frac{1}{f_0^2}\int_{-\infty}^{\infty} \gamma_{\delta f}(t)\exp(-j2\pi\nu t)\mathrm{d}t \quad (\mathrm{Hz}^{-1}) \tag{6.94}$$

for the continuous-time process $\delta f(t)$ and

$$p_{y,\tau}(\nu) = T\frac{1}{f_0^2}\sum_{-\infty}^{\infty} \gamma_{\delta f,\tau}(n)\exp(-j2\pi\nu nT) \quad (\mathrm{Hz}^{-1}) \tag{6.95}$$

for the discrete-time process $\delta f_\tau(n)$. T is the sampling time.

The spectral density of the reduced phase noise is

$$p_x(\nu) = \frac{1}{f_0^2}\int_{-\infty}^{\infty} \gamma_{\delta\phi}(t)\exp(-j2\pi\nu t)\mathrm{d}t \quad (\mathrm{Hz}^{-3}) \tag{6.96}$$

for the continuous-time process $\delta\phi(t)$ and

$$p_{x,\tau}(\nu) = T\frac{1}{f_0^2}\sum_{-\infty}^{\infty} \gamma_{\delta\phi,\tau}(n)\exp(-j2\pi\nu nT) \quad (\mathrm{Hz}^{-3}) \tag{6.97}$$

for the discrete-time process $\delta\phi_\tau(n)$. T is the sampling period.

In these expressions, $\gamma_{\delta f}(\tau)$ is the covariance of $\delta f(t)$, $\gamma_{\delta f,\tau}(n)$ is the covariance of $\delta f_\tau(n)$, $\gamma_{\delta\phi}(\tau)$ is the covariance of $\delta\phi(t)$ and $\gamma_{\delta\phi,\tau}(n)$ is the covariance of $\delta\phi_\tau(n)$.

According to (6.58) and (6.59), the spectral densities can also be computed from an estimator of the Fourier transform of the processes

$$p_{\delta f}(\nu) = \lim_{\Delta T\to\infty}\frac{1}{2\Delta T}\left|\int_{-\Delta T}^{\Delta T}\delta f(t)\exp(-j2\pi\nu t)\mathrm{d}t\right|^2 , \tag{6.98}$$

$$p_{\delta f,\tau}(\nu) = T\lim_{N\to\infty}\frac{1}{2N+1}\left|\sum_{-N}^{N}\delta f_\tau(n)\exp(-j2\pi\nu nT)\right|^2 . \tag{6.99}$$

The quantity

$$\int_{-\Delta T}^{\Delta T}\delta f(t)\exp(-j2\pi\nu t)\mathrm{d}t$$

is an estimator of the Fourier transform $\Delta F(\nu)$ of the continuous-time process $\delta f(t)$ and the quantity

$$\sum_{-N}^{N}\delta f_\tau(n)\exp(-j2\pi\nu nT)$$

is an estimator of the Fourier transform $\Delta F_\tau(\nu)$ of the discrete-time process $\delta f_\tau(n)$.

Since $x(t)$ and $y(t)$ are related by relations of (6.91) and (6.92), we have the simple relation between their two Fourier transforms $X(\nu)$ and $Y(\nu)$,

$$Y(\nu) = 2\pi j\nu X(\nu) \tag{6.100}$$

and between their spectral densities $p_x(\nu)$ and $p_y(\nu)$,

$$p_y(\nu) = 4\pi^2\nu^2 p_x(\nu) \; . \tag{6.101}$$

Model of the Frequency Noise of Ultra-stable Oscillators

The spectral density of the frequency noise of ultra-stable oscillators is frequently described by a superposition of components whose spectral densities are integer powers of the Fourier frequency ν [81].

In this model, there are five types of independent noise processes. The following power laws describe the five components of the dimensionless reduced frequency fluctuation, i.e. $y(t) = \delta f(t)/f_0$

$$p_{y,i}(\nu) = h_i \times |\nu|^i \; , \quad -2 \le i \le 2 \; . \tag{6.102}$$

The dimension of spectral density $p_{y,i}(\nu)$ of the dimensionless quantity $y(t)$ is Hz^{-1}.

These five components are described in Table 6.1.

Table 6.1. Noise processes involved in the frequency fluctuations of ultra-stable oscillators

Value of the power i	Dimension of h_i	Process
−2	Hz	Random walk frequency noise
−1	Dimensionless	Flicker frequency noise
0	Hz^{-1}	White frequency noise
+1	Hz^{-2}	Flicker phase noise
+2	Hz^{-3}	White phase noise

Random Walk Frequency Noise

A random walk is a simple stochastic process [128]. In the case of frequency noise, which is a 1D random walk, it describes successive frequency jumps, which may be positive or negative with the same probability.

It can be shown in this case [129] that the total variation of the frequency during a given time interval Δt asymptotes to $\pm\sqrt{2n/\pi} \approx \pm 0.8\sqrt{n}$.

This explains the strong growth of the spectral density for low frequencies, i.e. long sampling times.

Flicker Frequency Noise

Flicker noise, also known as $1/f$ noise or pink noise [132], is a process with a frequency spectrum such that the power spectral density is proportional to the reciprocal of the frequency.

Although it physical origin is not well elucidated, flicker noise is a process that appears in many physical situations. It is often dominant in the low Fourier frequency ranges.

Flicker Phase Noise

Equation (6.101) shows that the spectral density of the frequency fluctuations is proportional to that of the phase fluctuations times the square of the Fourier frequency,

$$\begin{aligned} p_{\mathrm{f}}(\nu) &\propto p_{\phi}(\nu) \times \nu^2 \\ &= h_1 \left|\nu\right| \ . \end{aligned} \tag{6.103}$$

Consequently, flicker phase noise produces a frequency noise component whose spectral density is proportional to the Fourier frequency ν.

White Frequency Noise

White noise occurs when there is no correlation between the values of the process for any two different times [135]. Thermal noise (Johnson noise), produced by the random thermal motion of the electrical charges in electronic circuits is the most known source of white frequency noise. Shot noise, which appears when discrete quanta are implied, e.g. electrons in an electronic circuit or photons in an optical experiment also produces white noise.

White Phase Noise

Similarly to the flicker phase noise, the white phase noise produces a frequency noise whose spectral density is proportional to that of the phase noise times the square of the Fourier frequency,

$$\begin{aligned} p_{\mathrm{f}}(\nu) &\propto p_{\phi}(\nu) \times \nu^2 \\ &= h_2 \nu^2 \ . \end{aligned} \tag{6.104}$$

6.2.3 Variance of the Frequency Fluctuations

The simplest statistical parameter that characterizes the random frequency fluctuations in the time domain is the variance of the dimensionless fluctuating part $\delta f(t)/f_0$ of the frequency. This variance is a measure of how spread out this random process is.

The input data is the discrete-time process $\delta f_\tau(n)$, which results from the frequency measurement and is related to the continuous-time process $\delta f(t)$ as described in Sect. 6.2.1.

Notice that the measurement is performed during a finite time interval and consequently only a limited sample of the discrete-time process is known. The variance of the discrete-time process must be first estimated from this finite sample, and then the nature of the frequency noise of the oscillator deduced from this estimation.

Notice that the algorithm includes the frequency drift, if any.

Estimator of the Variance

The first step is to characterize the quality of the estimator of the variance.

The Estimator s_x^2

When estimating the population variance using N samples $x(n)$ of the total population where $n = 1, 2, \cdots, N$, the following formula is used, which gives an *unbiased estimator* s_x^2 of the variance σ_x^2:

$$s_x^2 = \frac{1}{N-1} \sum_{n=1}^{N} \left(x(n) - \overline{x(n)} \right)^2 . \tag{6.105}$$

In this expression, the mean value $\overline{x(n)}$ is given by

$$\overline{x(n)} = \frac{1}{N} \sum_{p=1}^{N} x(p) . \tag{6.106}$$

An Unbiased Estimator

An unbiased estimator is one that, *on average*, does not overestimate or underestimate the quantity that is being estimated.

We note that

$$E\left\{ \sum_{n=1}^{N} \left(x(n) - \overline{x(n)} \right)^2 \right\} , \tag{6.107}$$

which is the average of the quantity used to calculate the estimator s_x^2. We can write

$$
\begin{aligned}
E\left\{\sum_{n=1}^{N}\left(x(n)-\overline{x(n)}\right)^2\right\} &= E\left\{\sum_{n=1}^{N}\left(x(n)-\mu-\left(\overline{x(n)}-\mu\right)\right)^2\right\} \\
&= \sum_{n=1}^{N} E\left\{\left((x(n)-\mu)-\left(\overline{x(n)}-\mu\right)\right)^2\right\} \\
&= \sum_{n=1}^{N} E\left\{(x(n)-\mu)^2\right\} \\
&\quad -2\sum_{n=1}^{N} E\left\{(x(n)-\mu)\left(\overline{x(n)}-\mu\right)\right\} \\
&\quad +\sum_{n=1}^{N} E\left\{\left(\overline{x(n)}-\mu\right)^2\right\} .
\end{aligned}
$$

In these expressions, μ is the (unknown) mean value of the sequence.

The first member of the right-hand side is the variance σ_x^2 of the sequence $x(n)$, multiplied by N,

$$
\begin{aligned}
\sum_{n=1}^{N} E\left\{(x(n)-\mu)^2\right\} &= \sum_{n=1}^{N} \sigma_x^2 \\
&= N \times \sigma_x^2 .
\end{aligned}
$$

To calculate the second and third members, we bear in mind that

$$
\overline{x(n)} = 1/N \sum_{m=1}^{N} x(m) .
$$

Consequently,

$$
\begin{aligned}
\sum_{n=1}^{N} E\left\{(x(n)-\mu)\left(\overline{x(n)}-\mu\right)\right\} &= \sum_{n=1}^{N} E\left\{\frac{1}{N}\sum_{m=1}^{N}(x(n)-\mu)(x(m)-\mu)\right\} \\
&= \sigma_x^2 ,
\end{aligned}
$$

$$
\begin{aligned}
\sum_{n=1}^{N} E\left\{\left(\overline{x(n)}-\mu\right)^2\right\} &= \sum_{n=1}^{N} E\left\{\left(\frac{1}{N}\sum_{m=1}^{N} x(m)-\mu\right)^2\right\} \\
&= \frac{1}{N^2}\sum_{n=1}^{N}\sum_{m=1}^{N} E\left\{(x(m)-\mu)^2\right\} \\
&= \sigma_x^2 .
\end{aligned}
$$

In conclusion, the average of the quantity used to calculate the estimator s_x^2 equals

$$E\left\{\sum_{n=1}^{N}\left(x(n)-\overline{x(n)}\right)^2\right\}=(N-1)\times\sigma_x^2\,.$$

Consequently, the quantity

$$s_x^2=\frac{\sum_{n=1}^{N}\left(x(n)-\overline{x(n)}\right)^2}{N-1}$$

is an unbiased estimator of the population variance of $x(n)$. This is called the sample variance of $x(n)$.

Information Given by the Variance

The notations are the following:

- instantaneous reduced frequency fluctuations $y(t)$ and their spectral density $p_y(\nu)$

$$y(t)=\frac{\delta f(t)}{f_0}\,, \tag{6.108}$$

- the continuous-time process $y_\tau(t)$ resulting from the frequency measurements over a duration τ and its spectral density $p_{y,\tau}(\nu)$

$$y_\tau(t)=\frac{\delta f_\tau(t)}{f_0}\,, \tag{6.109}$$

- the discrete-time process $y_\tau^T(n)$ resulting from sampling $y_\tau(t)$ with sampling rate $\nu_\mathrm{s}=1/T$ and its spectral density $P_{y,\tau}^T(\nu)$

$$y_\tau^T(n)=\frac{\delta f_\tau^T(n)}{f_0}\,. \tag{6.110}$$

What is called variance of the frequency fluctuations of an ultra-stable oscillator is in fact the variance of the discrete-time process $y_\tau^T(n)=\delta f_\tau^T(n)/f_0$. The sampling time T, equal to the integration time τ in the case where there is no dead time, which is the most frequent case, is the parameter of this quantity.

If $p(\nu)$ is the spectral density of the reduced frequency fluctuations $y(t)$ of the oscillator,

$$p_\tau(\nu)=p(\nu)\times\left|G_\mathrm{m}(\nu)\right|^2 \tag{6.111}$$

$$=p(\nu)\times\left[\frac{\sin\pi\nu\tau}{\pi\nu\tau}\right]^2 \tag{6.112}$$

is the power density of the continuous-time process $y_\tau(t)$ resulting from frequency measurements of duration τ.

The spectral density $P^T_{y,\tau}(\nu)$ of the discrete-time process $y^T_\tau(n)$ resulting from sampling $y_\tau(t)$ with the sampling rate $T = 1/\nu_\mathrm{s}$ is (see Sect. 6.2.2)

$$P^T_{y,\tau}(\nu) = \sum_{k=-\infty}^{\infty} p_\tau(\nu + k\nu_\mathrm{s}) \tag{6.113}$$

$$= \sum_{k=-\infty}^{\infty} p_y(\nu + k\nu_\mathrm{s}) \times [G_\mathrm{m}(\nu + k\nu_\mathrm{s})]^2 \tag{6.114}$$

$$= \sum_{k=-\infty}^{\infty} p_y(\nu + k\nu_\mathrm{s}) \times \left[\frac{\sin[\pi(\nu + k\nu_\mathrm{s})\tau]}{\pi(\nu + k\nu_\mathrm{s})\tau}\right]^2 . \tag{6.115}$$

The variance $\sigma^2_{y,T}(\tau)$ of the discrete-time process y^T_τ is consequently

$$\sigma^2_{\tau,T}(\tau) = \int_{-\nu_\mathrm{s}/2}^{\nu_\mathrm{s}/2} P^\mathrm{T}_{y,\tau}(\nu)\mathrm{d}\nu \tag{6.116}$$

$$= \int_{-\nu_\mathrm{s}/2}^{\nu_\mathrm{s}/2} \sum_{k=-\infty}^{\infty} p_y(\nu + k\nu_\mathrm{s}) \times \left[\frac{\sin[\pi(\nu + k\nu_\mathrm{s})\tau]}{\pi(\nu + k\nu_\mathrm{s})\tau}\right]^2 \mathrm{d}\nu . \tag{6.117}$$

Putting $f = \nu + k\nu_\mathrm{s}$

$$\sigma^2_{y,T}(\tau) = \sum_{k=-\infty}^{\infty} \int_{(k-1/2)\nu_s}^{(k+1/2)\nu_s} p_y(f) \times \left[\frac{\sin(\pi f \tau)}{\pi f \tau}\right]^2 \mathrm{d}f \tag{6.118}$$

$$= \int_{-\infty}^{\infty} p(f) \times \left[\frac{\sin(\pi f \tau)}{\pi f \tau}\right]^2 \mathrm{d}f . \tag{6.119}$$

Putting the dimensionless variable $x = \pi\tau\nu$

$$\sigma^2_{y,T}(\tau) = \frac{1}{\pi\tau} \int_{-\infty}^{\infty} \mathcal{P}(x) \frac{\sin^2(x)}{x^2} \mathrm{d}x , \tag{6.120}$$

where

$$\mathcal{P}(x) = p_y\left(\frac{x}{\pi\tau}\right) . \tag{6.121}$$

The result of the process of variance measurement is the sample variance $s^2_{y,T}(\tau)$ of the discrete-time process $y^\mathrm{T}_\tau(n)$. It is an unbiased estimator of the variance of this process. It is related to the variance of the instantaneous frequency fluctuations $y(t)$ but is different from it because of the filter associated to the frequency measurement process (see Sect. 6.2.1). It depends on the duration τ of each individual measurement and the effect of this experimental parameter must be studied.

The case $T = \tau$ (no dead time) is the most frequently encountered. In the following, this situation will be assumed. The notations in this case can be simplified and are as follows:

$$y_\tau^\tau(t) = y_\tau(t) \,, \tag{6.122}$$

$$P_{y,\tau}^\tau(\nu) = P_{y,\tau}(\nu) \,, \tag{6.123}$$

$$\sigma_{y,\tau}^2(\tau) = \sigma_y^2(\tau) \,. \tag{6.124}$$

Variance Values for the Components of the Frequency Noise

The law of the variation of $\sigma_y^2(\tau)$ as a function of the duration τ of each measurement is a characteristic of the noise that affects the frequency.

1. $p_y(\nu) = h_0$, $\mathcal{P}(x) = h_0$ (white frequency noise).
   In this case, we have (see (A.3))

$$\sigma_y^2(\tau) = h_0 \frac{1}{\pi\tau} \int_{-\infty}^{\infty} \left[\frac{\sin(x)}{x}\right]^2 \mathrm{d}x \tag{6.125}$$

$$= \frac{h_0}{\tau} \,. \tag{6.126}$$

   The variance of the filtered frequency fluctuations is inversely proportional to the duration τ of each measurement, since the filter width is inversely proportional to τ.
2. $p_y(\nu) = h_1 |\nu|$, $\mathcal{P}(x) = (h_1/\pi\tau)\,|x|$ (flicker phase noise).
   In this case, the finite width of the noise spectrum (or of the measurement apparatus) must be taken into account, otherwise the total power is infinite. $\nu_{\max}$ is the upper limit of the noise spectrum and $x_{\max} = \pi\tau\nu_{\max}$. We have (see (A.2))

$$\sigma_y^2(\tau) = 2\frac{h_{-1}}{\pi^2\tau^2} \int_0^{x_{\max}} \frac{\sin^2(x)}{x} \mathrm{d}x \tag{6.127}$$

$$= \frac{h_{-1}}{\pi^2\tau^2} \times [\ln(\pi\tau\nu_{\max}) + \gamma] \,. \tag{6.128}$$

   γ is the Euler–Mascheroni constant

$$\gamma = 0.577216\ldots \tag{6.129}$$

3. $p_y(\nu) = h_2\nu^2$, $\mathcal{P}(x) = (h_2/\pi^2\tau^2)x^2$ (white phase noise).
   In this case, the finite width of the noise spectrum (or of the measurement apparatus) must be taken into account, otherwise the total power is infinite. $\nu_{\max}$ is the upper limit of the noise spectrum and $x_{\max} = \pi\tau\nu_{\max}$. We have (see (A.1))

$$\sigma_y^2(\tau) = 2\frac{h_{-1}}{\pi^3\tau^3} \int_0^{x_{\max}} \sin^2(x)\mathrm{d}x \tag{6.130}$$

$$= \frac{h_{-1}}{\pi^3\tau^3} \times \left[\pi\tau\nu_{\max} - \frac{\sin(2\pi\tau\nu_{\max})}{2}\right] \,. \tag{6.131}$$

Assuming that $\pi\tau\nu_{\max} \gg 1$

$$\sigma_y^2(\tau) = \frac{h_{-1}}{\pi^2\tau^2} \times \nu_{\max} \,. \tag{6.132}$$

4. $p_y(\nu) = h_{-1}\left|\nu^{-1}\right|$, $\mathcal{P}(x) = (h_{-1} \times \pi\tau)\left|x^{-1}\right|$ (flicker frequency noise). In this case, the value of the low frequency cut-off of the power law is not defined and consequently the lower limit of the integration must be taken equal to zero.
We have

$$\sigma_y^2(\tau) = 2h_{-1}\int_0^{\infty} \frac{\sin^2(x)}{x^3}\mathrm{d}x \tag{6.133}$$

$$= \lim_{x\to 0} h_{-1} \times [1 - 2\gamma - 2\ln(x)] \tag{6.134}$$

$$= \infty \,. \tag{6.135}$$

Consequently, no calculation of the variance can be made if the low-frequency cut-off is not known.
5. $p_y(\nu) = h_{-2}\nu^{-2}$, $\mathcal{P}(x) = (h_{-2}\times\pi^2\tau^2)x^{-2}$ (random walk frequency noise). The conclusion is the same as in the previous case: the variance cannot be predicted.

The Effect of a Constant Frequency Drift

The concept of variance is not defined in the case of a non-stationary process since the mean value cannot be computed. The following question arises consequently in the case where a frequency drift exists: what is the meaning of the result given by this algorithm of the sample variance?

For instance, assuming a constant drift of the frequency during the sample variance measurement and $T = \tau$, we have

$$f(t) = f_0 + \delta f_0 + \delta f(t) + k \times t \,, \tag{6.136}$$

where $f(t)$ is the instantaneous frequency, f_0 is the nominal frequency, δf_0 is a constant offset due to inaccuracy, $\delta f(t)$ is the random frequency fluctuation with a null mean value, k is the frequency drift (in Hz $\times$ s^{-1}) and t is time.

Application of the sample variance algorithm to this oscillator gives the following results:

1. Mean frequency measurement over a duration τ.
The value $f_\tau(n)$ obtained at time n is

$$f_\tau(nT) = \frac{1}{\tau}\int_{(n-1)\tau}^{n\tau} f(t)\mathrm{d}t \tag{6.137}$$

$$= f_0 + \delta f_0 + k\tau\left(n - \frac{1}{2}\right) + \delta f_\tau(n\tau) \,. \tag{6.138}$$

2. Mean value of the N samples.
   The mean value of the N samples $f_\tau(n)$ is

$$\overline{f_\tau(n)} = \frac{1}{N}\sum_{n=1}^{N} f_\tau(n) \tag{6.139}$$

$$= \frac{1}{N}\sum_{n=1}^{N}\left\{f_0 + \delta f_0 + k\tau\left(n - \frac{1}{2}\right) + \delta f_\tau(n\tau)\right\} \tag{6.140}$$

$$= \frac{1}{N}\left[N f_0 + N\delta f_0 + k\tau\sum_{n=1}^{N}\left(n - \frac{1}{2}\right) + \sum_{n=1}^{N}\delta f_\tau(n\tau)\right] \tag{6.141}$$

$$= f_0 + \delta f_0 + k\tau\left(\frac{N}{2}\right) + \frac{1}{N}\sum_{n=1}^{N}\delta f_\tau(n\tau) \tag{6.142}$$

$$= f_0 + \delta f_0 + k\tau\left(\frac{N}{2}\right) . \tag{6.143}$$

   The mean value of $\delta f_\tau(n\tau)$ is null.
3. Value of the difference $f_\tau(n) - \overline{f_\tau(n)}$.
   The constant parts of the frequency cancel, but the constant drift leads to a term that cannot be distinguished from the term due to the random fluctuation

$$f_\tau(n) - \overline{f_\tau(n)} = \delta f_\tau(n) + k\tau\left(n - \frac{N+1}{2}\right) . \tag{6.144}$$

4. Value returned by the algorithm.
   The above expression shows that the constant frequency shift appears as an additional term $k\tau\left(n - \frac{N+1}{2}\right)$ superimposed to the random fluctuations $\delta f_\tau(n)$.
   Application of the algorithm consequently gives

$$S_y^2(\tau) = \frac{1}{f_0} \times \frac{1}{N-1}\sum_{n=1}^{N}\left[\delta f_\tau(n) + k\tau\left(n - \frac{N+1}{2}\right)\right]^2 \tag{6.145}$$

$$= s_y^2(\tau) \tag{6.146}$$

$$+ \frac{1}{f_0} \times \frac{1}{N-1} \times k^2\tau^2 \sum_{n=1}^{N}\left(n - \frac{N+1}{2}\right)^2$$

$$+ \frac{1}{f_0} \times \frac{2}{N-1} \times k\tau \sum_{n=1}^{N}\delta f_\tau(n)\left(n - \frac{N+1}{2}\right) .$$

   The first term is the sample variance of the frequency fluctuations, the two other terms are due to the constant frequency drift. The algorithm cannot separate the two contributions.

For instance, if N is an odd number $N = 2M - 1$,

$$S_y^2(\tau) = s_y^2(\tau) \tag{6.147}$$

$$+ \frac{1}{f_0} \times \frac{1}{N-1} \times k^2\tau^2 \sum_{n=1}^{N} (n-M)^2$$

$$+ \frac{1}{f_0} \times \frac{2}{N-1} \times k\tau \sum_{n=1}^{N} \delta f_\tau(n)\,(n-M)$$

$$= s_y^2(\tau) \tag{6.148}$$

$$+ \frac{1}{f_0} \times \frac{1}{N-1} \times k^2\tau^2 \sum_{n=1}^{M-1} p^2$$

$$+ \frac{1}{f_0} \times \frac{2}{N-1} \times k\tau \sum_{n=1}^{N} \delta f_\tau(n)(n-M)\,. \tag{6.149}$$

Since (see, for instance [43])

$$\sum_{1}^{M-1} p^2 = \frac{M(M-1)(2M-1)}{6}\,, \tag{6.150}$$

the second term equals

$$\frac{1}{f_0} \times k^2\tau^2 \frac{M(2M-1)}{6}\,. \tag{6.151}$$

The third term can be neglected since the mean value of $\delta f_\tau(n)$ is null. We then have

$$S_y^2(\tau) = s_y^2(\tau) \tag{6.152}$$

$$+ \frac{1}{f_0} \times k^2\tau^2 \frac{M(2M-1)}{6}\,.$$

The frequency drift appears like a frequency fluctuation that should lead to a variance proportional to τ^2 and depends strongly on the number N of samples; it is proportional to $N(N+1)/2$.

Variance is a very Imperfect Tool

These results show that the variance of the discrete-time process is a very imperfect tool.

1. No theoretical result can be obtained for some of the components of the noise that can affect the stability of the oscillator under test.
2. A frequency drift appears like a random fluctuation that should lead to a variance proportional to τ^2 and depends strongly on the number N of samples since it is proportional to $N(N+1)/2$.

6.2.4 Allan Variance

Allan variance, named after David W. Allan [4] (and also known as the two-sample variance) is the tool most frequently used to characterize the stability of high performance oscillators. It is defined as one half of the time average of the squares of the differences between successive readings of the frequency, each one being measured over a duration τ.

As for the variance, a dead time can be introduced between the measurements, leading to a value of the sampling time larger than the integration time.

The Allan variance $\sigma^2_{y,\mathrm{A}}$ of a discrete-time process $y_n = \delta f(n)/f_0$ is consequently defined by

$$\sigma^2_{y,\mathrm{A}}(\tau) = \frac{1}{2}\left\langle (y_{n+1} - y_n)^2 \right\rangle . \tag{6.153}$$

Allan Variance is a Filter

The algorithm of the calculation of the Allan variance of a discrete-time process y_n is the algorithm of

1. a digital filter applied to the process y_n, followed by
2. the calculation of the variance of the filtered process.

The notations are the same as previously: $y(t)$ is the continuous-time process $y(t) = \delta f(t)/f_0$, $y_\tau(t)$ the continuous-time process resulting from frequency measurements of duration τ and $y_\tau(n)$ the discrete-time process resulting from sampling this continuous-time process at times $t_n = nT$. $\nu_\mathrm{s} = 1/T$ is the sampling rate and $T = \tau$ (no dead time).

Let $h_\mathrm{A}(n)$ be the pulse response of a discrete-time linear filter defined by

$$\begin{cases} h_A(0) = 1/\sqrt{2}\,, \\ h_A(1) = -1/\sqrt{2}\,, \\ h_A(n) = 0 \;\forall\, n < 0,\; n > 1\,. \end{cases} \tag{6.154}$$

If the discrete-time process $y_\tau(n)$ is applied to the input of this filter, the output process $y_{\tau,A}(n)$ is

$$y_{\tau,A}(n) = \sum_{p=-\infty}^{\infty} h_\mathrm{A}(p) y(n-p) \tag{6.155}$$

$$= \frac{1}{\sqrt{2}}\left[(y_\tau(n) - y_\tau(n-1)\right] . \tag{6.156}$$

The mean value of this output process cancels

$$E\{y_{\tau,A}(n)\} = \lim_{N\to\infty} \frac{1}{2N} \sum_{-N+1}^{N} \frac{y_\tau(n) - y_\tau(n-1)}{\sqrt{2}} \tag{6.157}$$

$$= \lim_{N\to\infty} \frac{1}{2N\sqrt{2}} (y_\tau(N) - y_\tau(-N)) \tag{6.158}$$

$$= 0\ . \tag{6.159}$$

Its variance is, therefore,

$$\sigma^2_{y,A}(\tau) = E\left[y_{\tau,A}(n) - E\left[y_{\tau,A}(n)\right]\right]^2 \tag{6.160}$$

$$= E\left[y_{\tau,A}(n)\right]^2 \tag{6.161}$$

$$= \frac{1}{2} E\left[y_\tau(n) - y_\tau(n-1)\right]^2\ . \tag{6.162}$$

This is the definition of the Allan variance of the process $y_\tau(n)$.

Consequently, the Allan variance of a discrete-time random process $y_\tau(n)$ is the variance of the process resulting from filtering $y_\tau(n)$ with the filter whose impulse response is given by (6.154).

The frequency response of this filter is

$$H_A(\nu) = \sum_{n=-\infty}^{n=+\infty} h_A(n) \times \exp^{-j2n\pi\nu\tau} \tag{6.163}$$

$$= 1/\sqrt{2} \times (1 - \cos 2\pi\nu\tau + j \sin 2\pi\nu\tau)\ . \tag{6.164}$$

The square of its amplitude response is

$$|H_A(\nu)|^2 = \frac{1}{2}(1 - \cos 2\pi\nu\tau) \tag{6.165}$$

$$= \sin^2 \pi\nu\tau\ . \tag{6.166}$$

This filter is a band-pass filter whose amplitude response equals 1 for the frequency $1/2\tau = \nu_s/2$ and 0 for the null and the sampling frequencies. Its half power full width is $\delta\nu_{1/2} = \nu_s/2$. Figure 6.8 shows the squared amplitude response of the filter. It transmits without loss the Fourier components of the input signal near $1/2\tau$ and blocks the Fourier components of the input signal near the null frequency.

The signal at the output of this filter is consequently free of the null frequency component and contains mainly the Fourier components of the input signal near $1/2\tau$.

In conclusion:

- Allan variance $\sigma^2_{y,A}(\tau)$ gives an indication of the spectral density of the discrete-time process $y(n)$ near the Fourier frequency $1/2\tau = \nu_s/2$.
- Allan variance is not sensitive to the Fourier frequencies near 0.
- The process $y_\tau(n)$ results from filtering and sampling the continuous-time process $y(t)$, which is the process of interest.

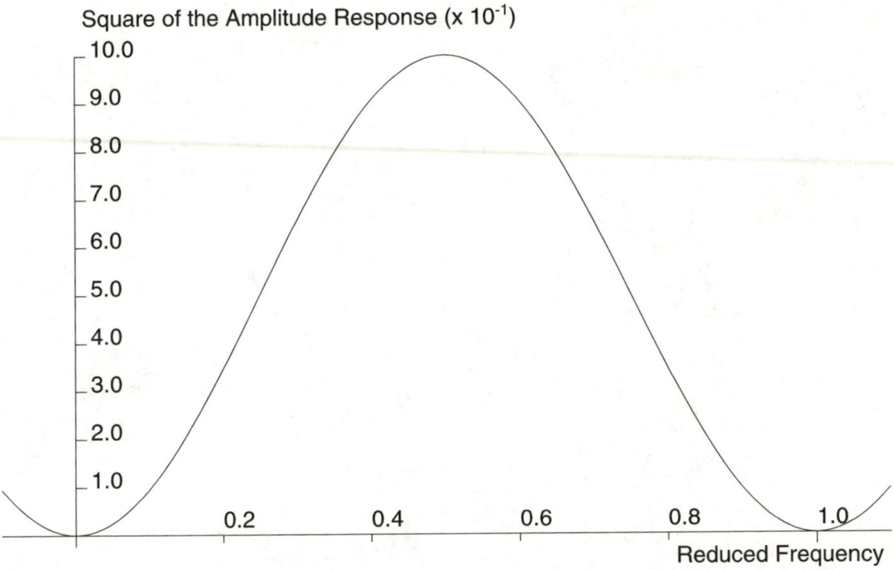

Fig. 6.8. Squared amplitude response of the Allan filter

Information Given by Allan Variance

Random Frequency Fluctuations

Allan variance is the variance of the discrete-time process $y_\tau(n)$ filtered by the filter $H_A(\nu)$ described in Sect. 6.2.4.

If $p_y(\nu)$ is the spectral density of the reduced frequency fluctuations $y(t)$,

$$p_{y,\tau}(\nu) = p_y(\nu) \times |G_m(\nu)|^2 \tag{6.167}$$

$$= p_y(\nu) \times \left[\frac{\sin \pi\nu\tau}{\pi\nu\tau}\right]^2 \tag{6.168}$$

is the power density of the continuous-time process $y_\tau(t)$ resulting from frequency measurements of duration τ.

The spectral density $P_{y,\tau}(\nu)$ of the discrete-time process $y_\tau(n)$ resulting from sampling $y_\tau(t)$ with the sampling rate $T = 1/\nu_s$ is (see Sect. 6.2.2)

$$P_{y,\tau}(\nu) = \sum_{k=-\infty}^{\infty} p_{y,\tau}(\nu + k\nu_s) \tag{6.169}$$

$$= \sum_{k=-\infty}^{\infty} p_y(\nu + k\nu_s) \times [G_m(\nu + k\nu_s)]^2 \tag{6.170}$$

$$= \sum_{k=-\infty}^{\infty} p_y(\nu + k\nu_s) \times \left[\frac{\sin[\pi(\nu + k\nu_s)\tau]}{\pi(\nu + k\nu_s)\tau}\right]^2 . \tag{6.171}$$

The variance $\sigma_y^2(\tau)$ of the discrete-time process $y_\tau(n)$ is then

$$\sigma_y^2(\tau) = \int_{-\nu_s/2}^{\nu_s/2} P_{y,\tau}(\nu)\mathrm{d}\nu \tag{6.172}$$

$$= \int_{-\nu_s/2}^{\nu_s/2} \sum_{k=-\infty}^{\infty} p_y(\nu + k\nu_s) \times \left[\frac{\sin\left[\pi(\nu + k\nu_s)\tau\right]}{\pi(\nu + k\nu_s)\tau}\right]^2 \mathrm{d}\nu \tag{6.173}$$

$$= \sum_{k=-\infty}^{\infty} \int_{(k-1/2)\nu_s}^{(k+1/2)\nu_s} p_y(f) \times \left[\frac{\sin(\pi f \tau)}{\pi f \tau}\right]^2 \mathrm{d}f \,. \tag{6.174}$$

The discrete-time process $y_\tau(n)$ is then filtered by the Allan filter $H_\mathrm{A}(\nu)$. The spectral density $P_{y,\tau,A}(\nu)$ of the resulting discrete-time process $y_{\tau,A}(n)$ is

$$P_{y,\tau,A}(\nu) = P_{y,\tau}(\nu) \times |H_\mathrm{A}(\nu)|^2 \tag{6.175}$$

$$= \sum_{k=-\infty}^{\infty} p_y(\nu + k\nu_s) \left[\frac{\sin\left[\pi(\nu + k\nu_s)\tau\right]}{\pi(\nu + k\nu_s)\tau}\right]^2 \sin^2(\pi\nu\tau) \,. \tag{6.176}$$

We supposed that there is no dead time; $1/\nu_s = T = \tau$.

Consequently,

$$\sin^2(\pi\nu\tau) = \sin^2\left[\pi(\nu + k\nu_s)\tau\right] \,. \tag{6.177}$$

The variance of this filtered discrete-time process, which is the Allan variance of the reduced frequency fluctuations, is then

$$\sigma_{y,A}^2(\tau) = \int_{-\nu_s/2}^{\nu_s/2} \sum_{k=-\infty}^{\infty} p_y(\nu + k\nu_s) \frac{\sin^4\left[\pi(\nu + k\nu_s)\tau\right]}{\left[\pi(\nu + k\nu_s)\tau\right]^2} \mathrm{d}\nu \tag{6.178}$$

$$= \int_{-\infty}^{\infty} p_y(f) \frac{\sin^4(\pi f \tau)}{(\pi f \tau)^2} \mathrm{d}f \,. \tag{6.179}$$

Putting the dimensionless variable $x = \pi\tau\nu$,

$$\sigma_{y,A}^2(\tau) = \frac{1}{\pi\tau} \int_{-\infty}^{\infty} \mathcal{P}(x) \frac{\sin^4(x)}{x^2} \mathrm{d}x \,, \tag{6.180}$$

where

$$\mathcal{P}(x) = p_y\left(\frac{x}{\pi\tau}\right) \,. \tag{6.181}$$

This can be detailed for the five components of the frequency noise.

1. $p_y(\nu) = h_0$, $\mathcal{P}(x) = h_0$ (white frequency noise). In this case, (A.14) gives

$$\sigma_{y,A}(\tau) = \frac{h_0}{2\tau} \,. \tag{6.182}$$

2. $p_y(\nu) = h_1 |\nu|$, $\mathcal{P}(x) = (h_1/\pi\tau)|x|$ (flicker phase noise).
   In this case, the width of the noise spectrum (or of the measurement apparatus) must be taken into account, otherwise the total power is infinite. $\nu_{\max}$ is the upper limit of the noise spectrum and $x_{\max} = \pi\tau\nu_{\max}$.
   The Allan variance is

$$\sigma_{y,A}(\tau) = \frac{2}{\pi^2\tau^2} \int_0^{x_{\max}} x \frac{\sin^4(x)}{x^2} \mathrm{d}x \, . \tag{6.183}$$

   Equation (A.15) gives

$$\sigma_{y,A}(\tau) = \frac{h_1}{4\pi^2\tau^2} \left[3.118 + 3\ln(\pi\nu_{\max}\tau)\right] \, . \tag{6.184}$$

3. $p_y(\nu) = h_2\nu^2$, $\mathcal{P}(x) = (h_2/\pi^2\tau^2)x^2$ (white phase noise).
   In this case, the width of the noise spectrum (or of the measurement apparatus) must be taken into account.
   The Allan variance is then given by (see (A.16))

$$\sigma_{y,A}(\tau) = \frac{2}{\pi^3\tau^3} \int_0^{x_{\max}} \sin^4(x)\mathrm{d}x \tag{6.185}$$

$$= h_2 \frac{3\nu_{\max}}{4\pi^2} \frac{1}{\tau^2} \, . \tag{6.186}$$

4. $p_y(\nu) = h_{-1} |\nu^{-1}|$, $\mathcal{P}(x) = (h_{-1} \times \pi\tau) |x^{-1}|$ (flicker frequency noise).
   The Allan variance is given by (see (A.17))

$$\sigma_{y,A}(\tau) = h_{-1} \int_{-\infty}^{\infty} \frac{\sin^4(x)}{|x^3|} \mathrm{d}x \tag{6.187}$$

$$= 2\ln(2) h_{-1} \, . \tag{6.188}$$

5. $p_y(\nu) = h_{-2}\nu^{-2}$, $\mathcal{P}(x) = (h_{-2} \times \pi^2\tau^2)x^{-2}$ (random walk frequency noise).
   The Allan variance is given by (see (A.18))

$$\sigma_{y,A}(\tau) = h_{-2}\pi\tau \int_{-\infty}^{\infty} \frac{\sin^4(x)}{x^4} \mathrm{d}x \tag{6.189}$$

$$= \frac{2\pi^2}{3} h_{-2} \times \tau \, . \tag{6.190}$$

Constant Frequency Drift

Assume a constant drift of the frequency during variance measurement

$$f(t) = f_0 + \delta f_0 + \delta f(t) + k \times t \, . \tag{6.191}$$

In this relation, $f(t)$ is the instantaneous frequency, f_0 is the nominal frequency, δf_0 is a constant offset due to inaccuracy, $\delta f(t)$ is a random frequency fluctuation k is the frequency drift and t is time.

Application of the Allan variance algorithm gives the following results:

- The value $f_\tau(n-1)$ obtained at time $(n-1)\tau$ is

$$f_\tau(n-1) = \frac{1}{\tau}\int_{(n-2)\tau}^{(n-1)\tau} f(t)\mathrm{d}t \tag{6.192}$$

$$= f_0 + \delta f_0 + \delta f(n-1) + k\tau\left(n-1-\frac{1}{2}\right) . \tag{6.193}$$

- The value $f_\tau(n)$ obtained at time $n\tau$ is

$$f_\tau(n) = \frac{1}{\tau}\int_{(n-1)\tau}^{n\tau} f(t)\mathrm{d}t \tag{6.194}$$

$$= f_0 + \delta f_0 + \delta f(n) + k\tau\left(n-\frac{1}{2}\right) . \tag{6.195}$$

- The difference is

$$f_\tau(n) - f_\tau(n-1) = \delta f(n) - \delta f(n-1) + k\tau . \tag{6.196}$$

- The value returned by the algorithm

$$S^2_{A,y}(\tau) = \frac{1}{2f_0^2}\frac{1}{N-1}\sum_1^N(\delta f(n) - \delta f(n-1) + k\tau)^2 \tag{6.197}$$

$$= \frac{1}{2f_0^2}\frac{1}{N-1}\sum_1^N(\delta f(n) - \delta f(n-1))^2$$

$$+ 2k\tau\frac{1}{2f_0^2}\frac{1}{N-1}\sum_1^N(\delta f(n) - \delta f(n-1))$$

$$+ \frac{1}{2f_0^2}\frac{N}{N-1}(k\tau)^2 . \tag{6.198}$$

Since

$$E\delta f(n) - \delta f(n-1) = 0 \tag{6.199}$$

we can write

$$S^2_y(\tau) = s^2_y + \frac{1}{2f_0^2}\frac{N}{N-1}k^2\tau^2 . \tag{6.200}$$

The above expression shows that, as for the true variance, the constant frequency shift appears as an additional term

$$s\prime^2_y = \frac{1}{2f_0^2}\frac{N}{N-1}(k\tau)^2 . \tag{6.201}$$

This effect is consequently much smaller than that obtained in the case of the true variance, in the ration $N(N+1)/6$: for $N = 100$, this ratio is larger than 10^3.

Moreover, the response of the Allan variance to a constant frequency drift is proportional to τ^2, which is a response different from the response of this variance to the five possible components of the frequency noise; it is consequently in principle possible to detect a constant frequency drift in the experimental results given by the Allan variance.

Uncertainty in the Estimation of Allan Variance

Any real measurement is an estimation of the population variance using N samples $x(n)$ of the total (infinite) population.

The estimator used is an unbiased one (see Sect. 6.2.3)

$$s_x^2 = \frac{1}{N-1} \sum_{n=1}^{N} \left(x(n) - \overline{x(n)} \right)^2 . \tag{6.202}$$

Allan variance, as well as any other variances described in this book, is the variance of a discrete-time process $z(n)$ which results from filtering the process $y_\tau(n)$ by a specific filter. Consequently, since the same algorithm is used to compute the estimator of the resulting variance, this estimator is unbiased.

It is possible to evaluate the uncertainty resulting from the use of a finite number N of samples $z(n)$ [116, 79, 80, 138].

The value $s_{x,N}^2$ of the estimator defined by (6.202) is itself a random process, the number N of measurements used to compute it being a parameter.

The variance of this quantity has been calculated in the case of Allan variance [79, 80, 81]. It is then possible, knowing the number N, to draw error bars on any frequency stability graph that uses the Allan deviation.

For large values of N, the fractional error ϵ varies inversely with the square root of N,

$$\epsilon = \frac{K}{\sqrt{N}} . \tag{6.203}$$

$0.7 < K < 1.1$ is a coefficient that depends mainly on the dominant contribution to the frequency noise and on the dead time [81].

6.2.5 Hadamard Variance

Many other numeric filters can be (and have been) constructed on the scheme of Allan variance, with the aim of obtaining a more accurate description of the frequency noise. A best selectivity and a total rejection of constant frequency drift are some of the required objectives.

Hadamard variance is based on the Hadamard transform [130], adapted by R. A. Baugh [12].

As well as the Allan variance, the Hadamard variance filters the discrete-time process resulting from the frequency measurements.

This filter is defined by its impulse response, which may involve a large number of samples. Each sample is the result of a frequency measurement over a time τ. As usual, the sampling time may be different from the measurement duration τ if there is a dead time. We limit our discussion to the case $T = \tau$.

As a spectral estimator, the Hadamard transform consequently has a higher resolution than the Allan variance, whose impulse response involves only two samples. In fact, Allan variance is a particular case of Hadamard variance.

For the purposes of time-domain frequency stability characterization, the most important advantage of the Hadamard variance is its reduced sensitivity to a linear frequency drift, making it particularly useful for the analysis of frequency standards that show a large frequency drift, such as rubidium atomic frequency standards.

The Hadamard Filter

In its simplest form, the pulse response $h_{\mathrm{H}}(n)$ of the Hadamard filter is defined by

$$h_{\mathrm{H}}(n) = \begin{cases} (-1)^n & 0 \leq n \leq 2M-1 \\ 0 & n < 0,\ n \geq 2M \end{cases} . \tag{6.204}$$

In the case of the modified Hadamard variance, each sample $h_{\mathrm{H}}(n)$ of the pulse response is multiplied by the binomial coefficient $\binom{2M-1}{n}$ [81, 12]

$$h_{\mathrm{H}}(n) = \begin{cases} (-1)^n \times \binom{2M-1}{n} & 0 \leq n \leq 2M-1 \\ 0 & n < 0,\ n \geq 2M \end{cases} . \tag{6.205}$$

In both cases, if the discrete-time process $y_\tau(n)$ is applied to the input of this filter, the output sequence $y_{\tau,\mathrm{H}}(n)$ is

$$y_{\tau,\mathrm{H}}(n) = \sum_{p=-\infty}^{\infty} h_{\mathrm{H}}(p) y_\tau(n-p) \tag{6.206}$$

$$= \sum_{p=0}^{2M-1} (-1)^p \times a_p y_\tau(n-p) . \tag{6.207}$$

a_p equals 1 in the simple case, $\binom{2M-1}{n}$ in the case of the modified Hadamard variance. The mean value of this output process is

$$E\{y_{\tau,\mathrm{H}}(n)\} = \lim_{N\to\infty} \frac{1}{2N} \sum_{-N+1}^{N} \sum_{p=0}^{2M-1} (-1)^p a_p y_\tau(n-p) \tag{6.208}$$

$$= 0 \tag{6.209}$$

in both the simple case and the modified one, since

$$\binom{2M-1}{n} = \binom{2M-1}{2M-1-n} . \tag{6.210}$$

Its variance is therefore

$$\sigma^2_{y,\mathrm{H}}(\tau) = E\left\{\left[y_{\tau,\mathrm{H}}(n) - E\left[y_{\tau,\mathrm{H}}(n)\right]\right]^2\right\} \tag{6.211}$$

$$= E\left[y_{\tau,\mathrm{H}}(n)\right]^2 \tag{6.212}$$

$$= E\left[\sum_{p=0}^{2M-1} (-1)^p \times a_p y_\tau(n-p)\right]^2 . \tag{6.213}$$

This defines the Hadamard variance of the process $y_\tau(n)$, which is consequently the variance of the process filtered by the filter whose impulse response is given by (6.204) or (6.205).

The frequency response of this filter is in the simple case

$$H_\mathrm{H}(\nu) = \sum_{n=-\infty}^{n=+\infty} h_\mathrm{H}(n) \times \exp^{-j2n\pi\nu\tau} \tag{6.214}$$

$$= \sum_{n=0}^{2M-1} (-1)^n \times \exp^{-j2n\pi\nu\tau} \tag{6.215}$$

$$= j\exp[-j\pi\nu(2M-1)T]\frac{\sin(2\pi\nu MT)}{\cos(\pi\nu T)} . \tag{6.216}$$

The square of its amplitude response is

$$|H_\mathrm{H}(\nu)|^2 = \left[\frac{\sin(2M\pi\nu T)}{\cos(\pi\nu T)}\right]^2 . \tag{6.217}$$

This filter is a band-pass filter whose maximal amplitude response equals $2M$ for the frequency $1/2\tau = \nu_\mathrm{s}/2$. The null and the sampling frequencies are blocked. The half power width of the filter is approximately given by

$$\delta\nu_\mathrm{H} = \nu_\mathrm{s}\frac{6}{\pi(8M^2-1)} . \tag{6.218}$$

Figure 6.9 shows the squared amplitude response of the filter divided by $(2M)^2$ for $M = 4$.

This filter

- is a band-pass filter, centered at $\nu = \nu_\mathrm{s}/2$, as is the Allan filter;
- removes the null frequency as well as the Allan filter; and
- can be made much more selective than the Allan filter by choosing a large value of M.

The result is qualitatively the same in the case of the modified Hadamard variance. Figure 6.10 shows its squared amplitude response in the case $M = 8$.

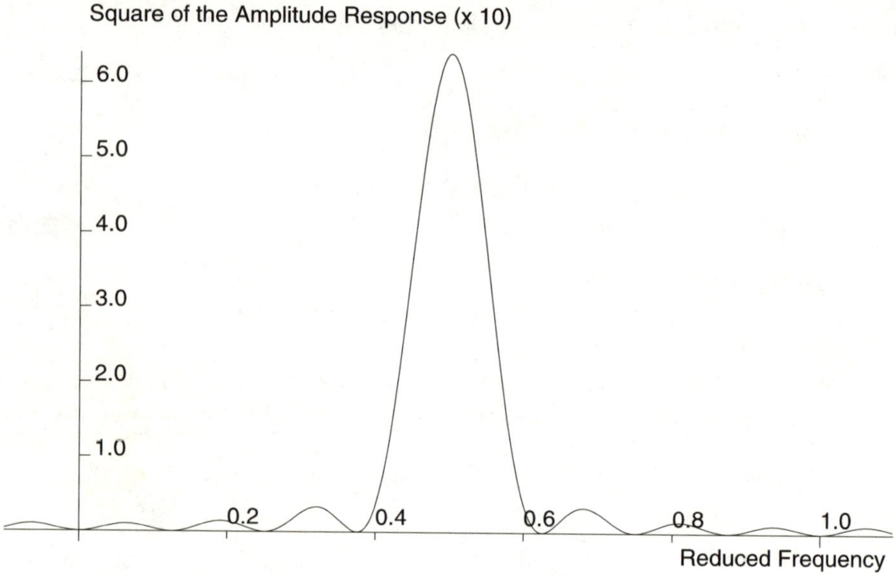

Fig. 6.9. Squared amplitude response of the Hadamard filter for $M = 4$

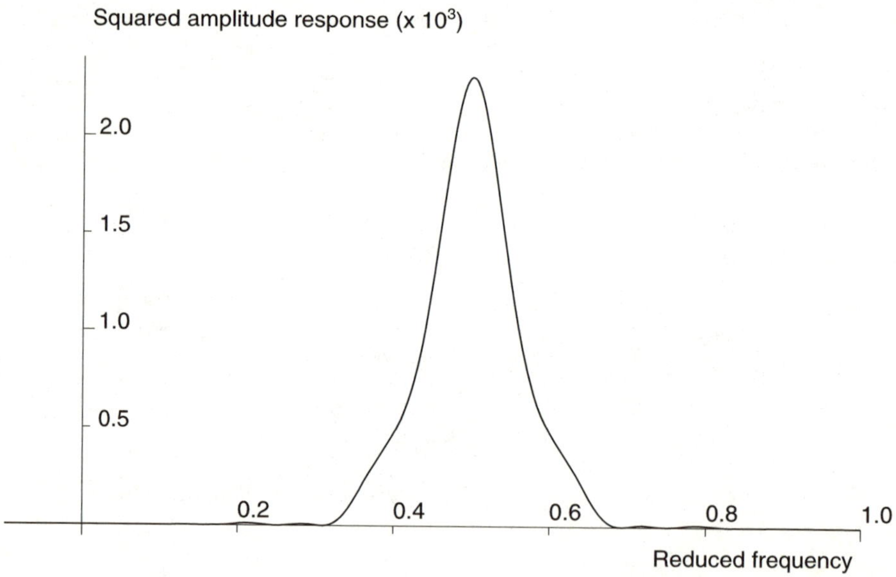

Fig. 6.10. Squared amplitude response of the modified Hadamard filter for $M = 8$

Information Given by a Hadamard Variance

Random Frequency Fluctuations

Hadamard variance is the variance of the discrete-time process $y_{\tau,H}(n)$, which results from filtering $y_\tau(n) = \delta f_\tau(n)/f_0$ by the filter $H_{\mathrm{H}}(\nu)$ described in Sect. 6.2.5.

The following discussion applies to the case of simple Hadamard variance, defined by (6.204). The results are qualitatively the same for the modified Hadamard variance.

If $p_y(\nu)$ is the spectral density of the reduced frequency fluctuations $y(t)$,

$$p_\tau(\nu) = p_y(\nu) \times |G_{\mathrm{m}}(\nu)|^2 \tag{6.219}$$

$$= p_y(\nu) \times \left[\frac{\sin \pi\nu\tau}{\pi\nu\tau}\right]^2 \tag{6.220}$$

is the power density of the continuous-time process $y_\tau(t)$ resulting from frequency measurements of duration τ.

The spectral density $P_\tau^{\mathrm{T}}(\nu)$ of the discrete-time process $y_\tau^{\mathrm{T}}(n)$ resulting from sampling $y_\tau(t)$ with the sampling rate $T = \tau = 1/\nu_{\mathrm{s}}$ (no dead time) is (see Sect. 6.2.2)

$$P_{y,\tau}(\nu) = \sum_{k=-\infty}^{\infty} p_\tau(\nu + k\nu_{\mathrm{s}}) \tag{6.221}$$

$$= \sum_{k=-\infty}^{\infty} p_\tau(\nu + k\nu_{\mathrm{s}}) \times [G_{\mathrm{m}}(\nu + k\nu_{\mathrm{s}})]^2 \tag{6.222}$$

$$= \sum_{k=-\infty}^{\infty} p_\tau(\nu + k\nu_{\mathrm{s}}) \times \left[\frac{\sin[\pi(\nu + k\nu_{\mathrm{s}})\tau]}{\pi(\nu + k\nu_{\mathrm{s}})\tau}\right]^2 . \tag{6.223}$$

The variance $\sigma_y^2(\tau)$ of the discrete-time process $y_\tau(n)$ is then

$$\sigma_y^2(\tau) = \int_{-\nu_{\mathrm{s}}/2}^{\nu_{\mathrm{s}}/2} P_{y,\tau}(\nu)\mathrm{d}\nu \tag{6.224}$$

$$= \int_{-\nu_{\mathrm{s}}/2}^{\nu_{\mathrm{s}}/2} \sum_{k=-\infty}^{\infty} p_y(\nu + k\nu_{\mathrm{s}}) \times \left[\frac{\sin[\pi(\nu + k\nu_{\mathrm{s}})\tau]}{\pi(\nu + k\nu_{\mathrm{s}})\tau}\right]^2 \mathrm{d}\nu \tag{6.225}$$

$$= \sum_{k=-\infty}^{\infty} \int_{(k-1/2)\nu_{\mathrm{s}}}^{(k+1/2)\nu_{\mathrm{s}}} p_y(f) \times \left[\frac{\sin(\pi f\tau)}{\pi f\tau}\right]^2 \mathrm{d}f . \tag{6.226}$$

The discrete-time process $y_\tau(n)$ is then filtered by $H_{\mathrm{H}}(\nu)$. The spectral density $P_{H,\tau}(\nu)$ of the resulting discrete-time process $y_{H,\tau}(n)$ is

$$P_{H,\tau}(\nu) = P_{y,\tau}(\nu) \times |H_{\mathrm{H}}(\nu)|^2$$

$$= \sum_{k=-\infty}^{\infty} p_y(\nu + k\nu_{\mathrm{s}}) \times \left[\frac{\sin[\pi(\nu + k\nu_{\mathrm{s}})\tau]}{\pi(\nu + k\nu_{\mathrm{s}})\tau}\right]^2 \times \left[\frac{\sin(2M\pi\nu\tau)}{\cos(\pi\nu\tau)}\right]^2 . \tag{6.227}$$

Since $\nu_\mathrm{s} = 1/T = 1/\tau$, we have

$$\sin^2(2M\pi\nu\tau) = \sin^2[2M\pi(\nu + k\nu_\mathrm{s})\tau] \ , \tag{6.228}$$

$$\cos^2(\pi\nu T) = \cos^2(\pi(\nu + k\nu_\mathrm{s})\tau) \ . \tag{6.229}$$

The variance $\sigma^2_{y,H}(\tau)$ of this discrete-time process, *which is the Hadamard variance of the reduced frequency fluctuations* is consequently

$$\sigma^2_{y,H}(\tau) = \sum_{k=-\infty}^{\infty} \int_{(k-1/2)\nu_\mathrm{s}}^{(k+1/2)\nu_\mathrm{s}} p(\nu + k\nu_\mathrm{s}) \times \frac{\sin^2[\pi(\nu + k\nu_\mathrm{s})\tau]}{[\pi(\nu + k\nu_\mathrm{s})\tau]^2} \times \frac{\sin^2(2M\pi(\nu + k\nu_\mathrm{s})\tau)}{\cos^2(\pi(\nu + k\nu_\mathrm{s})\tau)} \mathrm{d}\nu \ . \tag{6.230}$$

With $f = \nu + k\nu_\mathrm{s}$

$$\sigma^2_{y,H}(\tau) = \int_{-\infty}^{\infty} p(f) \frac{\sin^2(\pi f \tau)}{(\pi f \tau)^2} \left[\frac{\sin(2M\pi f\tau)}{\cos(\pi f \tau)}\right]^2 \mathrm{d}f \ . \tag{6.231}$$

Putting the dimensionless variable $x = \pi\tau\nu$

$$\sigma^2_{y,H}(\tau) = \frac{1}{\pi\tau} \int_{-\infty}^{\infty} \mathcal{P}(x) \frac{\sin^2(x)}{x^2} \frac{\sin^2(2Mx)}{\cos^2(x)} \mathrm{d}x \ . \tag{6.232}$$

In this equation,

$$\mathcal{P}(x) = p\left(\frac{x}{\pi\tau}\right) \ . \tag{6.233}$$

This can be detailed for the five power laws of the frequency noise model.

The results are given for $2M = 6$.

1. $p_y(\nu) = h_0$, $\mathcal{P}(x) = h_0$ (white frequency noise).
   In this case, (A.20) gives

$$\sigma_{y,H}(\tau) = \frac{h_0}{\pi\tau} \int_{-\infty}^{\infty} \frac{\sin^2(x)}{x^2} \frac{\sin^2(6x)}{\cos^2(x)} \mathrm{d}x \tag{6.234}$$

$$= \frac{4h_0}{\tau} \ . \tag{6.235}$$

2. $p_y(\nu) = h_1 |\nu|$, $\mathcal{P}(x) = (h_1/\pi\tau)|x|$ (flicker phase noise).
   In this case, the width of the noise spectrum (or of the measurement apparatus) must be taken into account, otherwise the total power is infinite. ν_{max} is the upper limit of the noise spectrum and $x_\mathrm{max} = \pi\tau\nu_\mathrm{max}$.
   The Hadamard variance is then given by (see (A.22))

$$\sigma_{y,H}(\tau) = \frac{2h_1}{(\pi\tau)^2} \int_0^{x_\mathrm{max}} \frac{\sin^2(x)}{x} \frac{\sin^2(6x)}{\cos^2(x)} \mathrm{d}x \tag{6.236}$$

$$= \frac{11h_1}{\pi^2} [\ln(\pi\nu_\mathrm{max}\tau) + \gamma] \frac{1}{\tau^2} \ . \tag{6.237}$$

3. $p_y(\nu) = h_2\nu^2$, $\mathcal{P}(x) = (h_2/\pi^2\tau^2)x^2$ (white phase noise).
   In this case, the width of the noise spectrum (or of the measurement apparatus) must be taken into account, otherwise the total power is infinite. $\nu_{\max}$ is the upper limit of the noise spectrum and $x_{\max} = \pi\tau\nu_{\max}$.
   The Hadamard variance is then given by (see (A.24))

$$\sigma_{y,H}(\tau) = \frac{2h_2}{(\pi\tau)^3}\int_0^{x_{\max}} \sin^2(x)\frac{\sin^2(6x)}{\cos^2(x)}\mathrm{d}x \tag{6.238}$$

$$= h_2\frac{7\nu_{\max}}{\pi^2}\frac{1}{\tau^2} \, . \tag{6.239}$$

4. $p_y(\nu) = h_{-1}\left|\nu^{-1}\right|$, $\mathcal{P}(x) = (h_{-1} \times \pi\tau)\left|x^{-1}\right|$ (flicker frequency noise).
   The Hadamard variance is given by (see (A.26))

$$\sigma_{y,H}(\tau) = h_{-1}\int_0^{\infty} \frac{\sin^2(x)}{x^3}\frac{\sin^2(6x)}{\cos^2(x)}\mathrm{d}x \tag{6.240}$$

$$= 13.42 \times h_{-1} \, . \tag{6.241}$$

5. $p_y(\nu) = h_{-2}\nu^{-2}$, $\mathcal{P}(x) = (h_{-2}\times\pi^2\tau^2)x^{-2}$ (random walk frequency noise).

   The Hadamard variance is given by (see (A.28))

$$\sigma_{y,H}(\tau) = h_{-2}\pi\tau\int_{-\infty}^{\infty} \frac{\sin^2(x)}{x^4}\frac{\sin^2(6x)}{\cos^2(x)}\mathrm{d}x \tag{6.242}$$

$$= 8\pi^2 h_{-2} \times \tau \, . \tag{6.243}$$

Constant Frequency Drift

Assume a constant drift of the frequency during variance measurement

$$f(t) = f_0 + \delta f_0 + \delta f(t) + k \times t \, . \tag{6.244}$$

In this relation, $f(t)$ is the instantaneous frequency, f_0 is the nominal frequency, δf_0 is a constant offset due to inaccuracy, $\delta f(t)$ is a random frequency fluctuation, k is the frequency drift coefficient and t is time.

Application of the simple Hadamard algorithm to this oscillator gives the following results (in the case $T = \tau$, no dead time).

– The value $f_\tau(n)$ obtained at time $(n)\tau$ is

$$f_\tau(n) = \frac{1}{\tau}\int_{(n-1)\tau}^{(n)\tau} f(t)\mathrm{d}t \tag{6.245}$$

$$= f_0 + \delta f_0 + \delta f(n) + k\tau\left(n - \frac{1}{2}\right) \, . \tag{6.246}$$

- The value $f_\tau(n-p)$ obtained at time $(n-p)\tau$ is

$$f_\tau(n-p) = \frac{1}{\tau}\int_{(n-p-1)\tau}^{(n-p)\tau} f(t)\mathrm{d}t \tag{6.247}$$

$$= f_0 + \delta f_0 + \delta f(n-p) + k\tau\left(n-p-\frac{1}{2}\right) . \tag{6.248}$$

- The output $f_{H,\tau}^{\mathcal{T}}(n)$ of the filter is

$$\begin{aligned} f_{H,\tau}^{\mathcal{T}}(n) &= \sum_{p=0}^{2M-1}(-1)^p f_\tau(n-p) \\ &= \sum_{p=0}^{2M-1}(-1)^p \delta f(n-p) + k\tau\sum_{p=0}^{2M-1}(-1)^p\left(n-p-\frac{1}{2}\right) \\ &= \sum_{p=0}^{2M-1}(-1)^p \delta f(n-p) + k\tau\sum_{p=0}^{2M-1}(-1)^p(-p) . \end{aligned} \tag{6.249}$$

However, we have

$$\sum_{p=0}^{2M-1}(-1)^p(-p) = M \tag{6.250}$$

and consequently

$$f_{H,\tau}(n) = \sum_{p=0}^{2M-1}(-1)^p \delta f(n-p) + Mk\tau . \tag{6.251}$$

- The value $S_{H,y}^2(\tau)$ returned by the algorithm applied to N frequency measurements is

$$S_{H,y}^2(\tau) = \frac{1}{f_0^2}\frac{1}{N-1}\sum_1^N \left(f_{H,\tau}(n)\right)^2 \tag{6.252}$$

$$\begin{aligned} &= \frac{1}{f_0^2}\frac{1}{N-1}\sum_1^N\left(\sum_{p=0}^{2M-1}(-1)^p\delta f(n-p) + Mk\tau\right)^2 \\ &= \frac{1}{f_0^2}\frac{1}{N-1}\sum_1^N\left(\sum_{p=0}^{2M-1}(-1)^p\delta f(n-p)\right)^2 \\ &\quad + \frac{1}{f_0^2}\frac{N}{N-1}M^2k^2\tau^2 \\ &\quad + \frac{2}{f_0^2}Mk\tau\frac{1}{N-1}\sum_1^N\left(\sum_{p=0}^{2M-1}(-1)^p\delta f(n-p)\right) . \end{aligned} \tag{6.253}$$

Since

$$E\delta f(n) = 0 \,, \tag{6.254}$$

we can write

$$\begin{aligned} S^2_{H,y}(\tau) &= \frac{1}{f_0^2}\frac{1}{N-1}\sum_1^N \left(\sum_{p=0}^{2M-1}(-1)^p \delta f(n-p)\right)^2 \\ &+ \frac{1}{f_0^2}\frac{N}{N-1}M^2k^2\tau^2 \\ &= \sigma^2_{H,y}(\tau) \\ &+ \frac{1}{f_0^2}\frac{N}{N-1}M^2k^2\tau^2 \,. \end{aligned} \tag{6.255}$$

In the above expression, $\sigma^2_{H,y}(\tau)$ is the result that should have been obtained if there had been no drift and

$$\frac{1}{f_0^2}\frac{N}{N-1}M^2k^2\tau^2 \tag{6.256}$$

is the additional term due to the drift.

This result shows that, as for the true variance and the Allan variance, the constant frequency shift appears as an additional term.

This additional term is much smaller the the effect obtained in the case of the true variance, in the ration $N(N+1)/6$. For $N = 100$, this ratio is larger than 10^3.

Moreover, as in the case of Allan variance, the response of the Hadamard variance to a constant frequency drift is proportional to τ^2, which is a response different from the response of this variance to the five possible components of the frequency noise; it is consequently in principle possible to detect a constant frequency drift in the experimental results given by the Hadamard variance.

6.2.6 Getting Rid of a Constant Frequency Drift

For both Allan and Hadamard variances, a constant frequency drift superimposed to a stationary random frequency fluctuation produces an additional term that cannot be directly separated from the one due to the random frequency fluctuations.

This is due to the fact that both algorithms are not balanced in order to cancel the effect of the drift for successive measurements of duration τ.

- In the Allan variance algorithm, the two successive measurements of the mean value of the drift do not give the same result; their difference does not cancel. The drift contribution is consequently $k\tau$, if k is the slope of the drift.

– In the Hadamard algorithm, this effect is cumulated on the $2M$ successive measurements, giving a contribution of $Nk\tau$.

A simple way to design an algorithm that cancels the effect of the frequency drift is to superpose two similar unbalanced algorithms, one giving a positive contribution of the drift and the other giving a negative one that exactly cancels the first.

Modifying the Hadamard Algorithm

Equation (6.249) shows that the effect of the drift is given by the sum

$$\sum_{p=0}^{2M-1} (-1)^p(-p) = M . \tag{6.257}$$

If the Hadamard filter is modified according to the new impulse response $h_{Hm}(n)$,

$$\begin{cases} h_{Hm}(0) = 1 \\ h_{Hm}(1) = -1 \\ \dots \\ h_{Hm}(2M-2) = 1 \\ h_{Hm}(2M-1) = -2 \\ h_{Hm}(2M) = 1 \\ \dots \\ h_{Hm}(4M-2) = 1 \\ h_{Hm}(n) = 0 \ \forall n < 0, \ n > 4M-2 . \end{cases} \tag{6.258}$$

The sum

$$\begin{aligned} \sum_{p=0}^{4M-2} h_{Hm}(n)(-p) &= \sum_{p=0}^{2M-1} (-1)^p(-p) + \sum_{p=2M-1}^{4M-2} (-1)^p(-p) \\ &= M - M \\ &= 0 \end{aligned} \tag{6.259}$$

and any *linear* frequency drift is canceled by the algorithm.

The frequency response of this balanced Hadamard filter is

$$\begin{aligned} H_{Hm} = (\nu) &= \sum_{n=0}^{2M-1} (-1)^n \times \exp^{-j2n\pi\nu\tau} + \sum_{n=2M-1}^{4M-2} (-1)^n \times \exp^{-j2n\pi\nu\tau} \\ &= j\exp[-j(2M-1)\pi\nu T]\frac{\sin(2M\pi\nu T)}{\cos(\pi\nu T)} \\ &\quad \times \left[1+(-1)^{2M-1}\exp[-j(2M-1)2\pi\nu T]\right] \\ &= -2\exp[-j(2M-1)2\pi\nu T]\frac{\sin(2M\pi\nu T)\sin[(2M-1)\pi\nu T]}{\cos(\pi\nu T)} \\ &= -\exp[-j(2M-1)2\pi\nu T]\frac{\cos(\pi\nu T)-\cos[(4M-1)\pi\nu T]}{\cos(\pi\nu T)} \\ &= -\exp[-j(2M-1)2\pi\nu T]\left\{1-\frac{\cos[(4M-1)\pi\nu T]}{\cos(\pi\nu T)}\right\} . \end{aligned} \tag{6.260}$$

Its squared amplitude response is

$$|H_{Hm}|^2 = \left\{1-\frac{\cos[(4M-1)\pi\nu T]}{\cos(\pi\nu T)}\right\}^2 , \tag{6.261}$$

which is very close to the response of the classical Hadamard variance using nearly the same number of samples (in fact, there is a difference of one sample), as shown in Fig. 6.11.

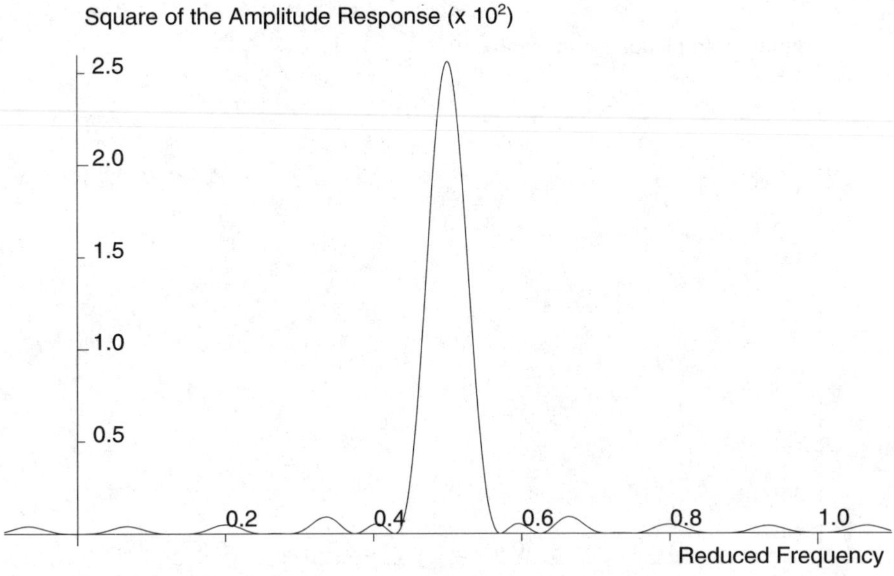

Fig. 6.11. Squared amplitude response of the balanced Hadamard filter with $(M = 4)$

The Simplest Algorithm: Three-samples Variance

This is the algorithm described above with $M = 1$. In this case, the impulse response of the filter is

$$\begin{cases} h_3(0) = 1 \\ h_3(1) = -2 \\ h_3(2) = 1 \\ h_3(n) = 0 \; \forall n < 0, \; n > 2 \, . \end{cases} \tag{6.262}$$

This is the so-called three-samples variance, introduced by Boileau and Picinbono [16]. The results of (6.261) with $M = 1$ give for its squared amplitude response

$$\begin{aligned} |H_3|^2 &= \left\{ 1 - \frac{\cos[(3)\pi\nu T]}{\cos(\pi\nu T)} \right\}^2 \\ &= 16 \sin^4(\pi\nu T) \, . \end{aligned} \tag{6.263}$$

Figure 6.12 shows the squared amplitude response of this filter.

The value of the three-samples variance $\sigma^2_{y,3}(\tau)$ for a frequency noise whose power spectra is $p(\nu)$ is given by (assuming $T = \tau$, no dead time)

$$\sigma^2_{y,3}(\tau) = 16 \int_{-\infty}^{\infty} p(\nu) \frac{\sin^2(\pi\nu\tau)}{(\pi\nu\tau)^2} \sin^4(\pi\nu\tau) \mathrm{d}\nu \, . \tag{6.264}$$

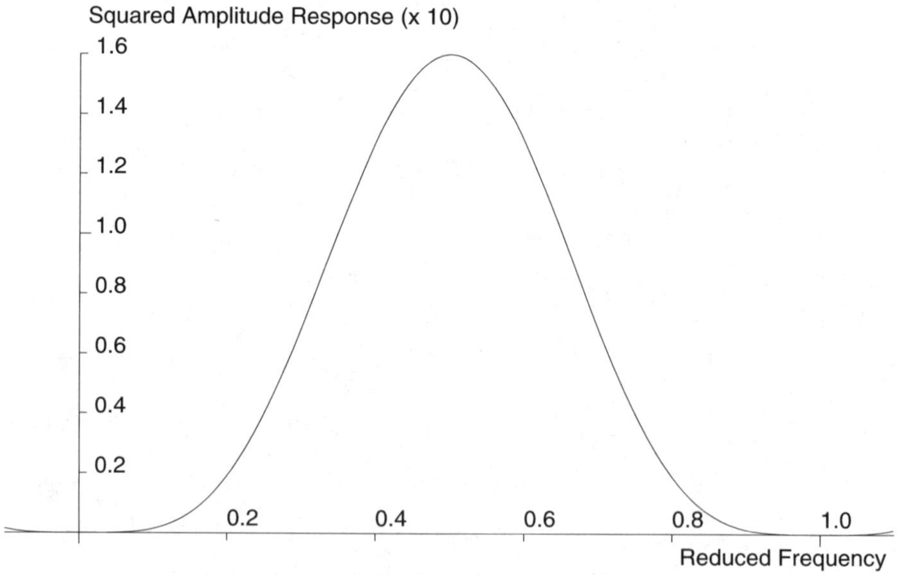

Fig. 6.12. Squared amplitude response of the three-samples filter

Using the reduced variable $x = \pi\nu\tau$, the three-samples variance can be written as

$$\sigma_{y,3}^2(\tau) = \frac{16}{\pi\tau}\int_{-\infty}^{\infty} \mathcal{P}(x)\frac{\sin^6(x)}{(x)^2}\mathrm{d}x \tag{6.265}$$

with

$$\mathcal{P}(x) = p\left(\frac{x}{\pi\tau}\right) . \tag{6.266}$$

This can be detailed for the various power laws of the frequency noise model.

1. $p_y(\nu) = h_0$, $\mathcal{P}(x) = h_0$ (white frequency noise). In this case, (A.31) gives

$$\sigma_{y,3}(\tau) = \frac{6h_0}{\tau} . \tag{6.267}$$

2. $p_y(\nu) = h_1\,|\nu|$, $\mathcal{P}(x) = (h_1/\pi\tau)\,|x|$ (flicker phase noise).
In this case, the width of the noise spectrum (or of the measurement apparatus) must be taken into account, otherwise the total power is infinite. $\nu_{\max}$ is the upper limit of the noise spectrum and $x_{\max} = \pi\tau\nu_{\max}$.
The three-samples variance is then given by (see (A.30))

$$\sigma_{y,3}(\tau) = \frac{32h_1}{(\pi\tau)^2}\int_0^{x_{\max}} \frac{\sin^6(x)}{x}\mathrm{d}x \tag{6.268}$$

$$= \frac{10h_1}{\pi^2}\left[\ln(\pi\nu_{\max}\tau) + \gamma\right]\frac{1}{\tau^2} . \tag{6.269}$$

3. $p_y(\nu) = h_2\nu^2$, $\mathcal{P}(x) = (h_2/\pi^2\tau^2)x^2$ (white phase noise).
In this case, the width of the noise spectrum (or of the measurement apparatus) must be taken into account, otherwise the total power is infinite. $\nu_{\max}$ is the upper limit of the noise spectrum and $x_{\max} = \pi\tau\nu_{\max}$.
The three-samples variance is then given by (see (A.29))

$$\sigma_{y,3}(\tau) = \frac{32h_2}{(\pi\tau)^3}\int_0^{x_{\max}} \sin^6(x)\mathrm{d}x \tag{6.270}$$

$$= h_2\frac{10\nu_{\max}}{\pi^2}\frac{1}{\tau^2} . \tag{6.271}$$

4. $p_y(\nu) = h_{-1}\left|\nu^{-1}\right|$, $\mathcal{P}(x) = (h_{-1}\times\pi\tau)\left|x^{-1}\right|$ (flicker frequency noise).
The three-samples variance is given by (see (A.32))

$$\sigma_{y,H}(\tau) = 32h_{-1}\int_0^{\infty} \frac{\sin^6(x)}{x^3}\mathrm{d}x \tag{6.272}$$

$$= 13.50 \times h_{-1} . \tag{6.273}$$

5. $p_y(\nu) = h_{-2}\nu^{-2}$, $\mathcal{P}(x) = (h_{-2}\times\pi^2\tau^2)x^{-2}$ (random walk frequency noise).

The three-samples variance is given by (see (A.33))

$$\sigma_{y,H}(\tau) = 32\pi\tau h_{-2}\int_0^\infty \frac{\sin^6(x)}{x^4}\mathrm{d}x \tag{6.274}$$

$$= 128\pi^2 h_{-2} \times \tau\ . \tag{6.275}$$

This simple algorithm gives results quite similar to those of Allan and Hadamard variances and is *free from any constant frequency drift.*

6.2.7 The Best Variance for a Given Application

Many kinds of variances have been proposed and their properties are slightly different. Facing a given application, the question arises as to the choice of the best algorithm for this application.

The following discussion is focused on three variances:

- Allan variance,
- Hadamard variance (with $2M = 6$), and
- three-samples variance.

Summary of the Results Given by Allan, Hadamard and Three-samples Variances

Table 6.2 summarizes the values of the bandwidth of the variances.

The bandwidth of the Hadamard variance can be made very small by choosing large values of the number $2M$ of samples used in its algorithm.

The bandwidth of Allan and three-samples variances are very similar.

Table 6.3 shows that the responses of the three variances are very similar; this is not a criterion for choosing a variance.

In fact, the main criteria for choosing a variance can hardly be predicted from these results. They are as follows:

- The ability to give a result free from frequency drifts. Only the three-samples variance possesses this property.
- The ability to characterize the frequency or phase noise that perturbs the frequency of the oscillator under test. This will be detailed in the next section.

Table 6.2. Bandwidth of Allan, Hadamard and three-samples variances

Variance	Bandwidth
Allan	$0.250 \times \nu_s$
Hadamard ($2M = 6$)	$0.070 \times \nu_s$
Three-samples	$0.182 \times \nu_s$

Table 6.3. Response to some noise spectral density laws

Power law	Variance	Result
	Allan	$h_{-2} \times 2\pi^2/3 \times \tau^{+1}$
$h_{-2}\nu^{-2}$	Hadamard ($2M = 6$)	$h_{-2} \times 8\pi^2 \times \tau^{+1}$
	Three-samples	$h_{-2} \times 128\pi^2 \times \tau^{+1}$
	Allan	$h_{-1} \times 1.39 \times \tau^0$
$h_{-1}\,\lvert\nu\rvert^{-1}$	Hadamard ($2M = 6$)	$h_{-1} \times 13.42 \times \tau^0$
	Three-samples	$h_{-1} \times 13.50 \times \tau^0$
	Allan	$h_0/2 \times \tau^{-1}$
$h_0\nu^0$	Hadamard ($2M = 6$)	$h_0 \times 4 \times \tau^{-1}$
	Three-samples	$h_0 \times 6 \times \tau^{-1}$
	Allan	$h_1 \times 3\left[\ln(\pi\nu_{\max}\tau) + 1.04\right]/4\pi^2 \times \tau^{-2}$
$h_1\,\lvert\nu\rvert^{+1}$	Hadamard ($2M = 6$)	$h_1 \times 11\left[\ln(\pi\nu_{\max}\tau) + 0.56\right]/\pi^2 \times \tau^{-2}$
	Three-samples	$h_1 \times 10\left[\ln(\pi\nu_{\max}\tau) + 0.56\right]/\pi^2 \times \tau^{-2}$
	Allan	$h_2 \times 3\nu_{\max}/4\pi^2 \times \tau^{-2}$
$h_2\nu^{+2}$	Hadamard ($2M = 6$)	$h_2 \times 7\nu_{\max}/\pi^2 \times \tau^{-2}$
	Three-samples	$h_2 \times 10\nu_{\max}/\pi^2 \times \tau^{-2}$

Analysis of the Frequency Noise

The frequency noise that perturbs the output frequency of an ultra-stable oscillator is a combination of the various power laws described in Sect. 6.2.2.

For a given Fourier frequency, one of these components is predominant.

The ability of the different variances to identify which component is dominant in a given Fourier frequency range is its main quality; this will be studied now.

Figure 6.13 shows the result of an ideal superposition of the five power laws. Coefficients h_2, h_1, h_0, h_{-1} and h_{-2} of the power laws have been chosen in such a way that every component is predominant in a range of frequencies of five decades. The absolute values of frequencies and spectral densities are of no importance for this abstract example.

The components are the following:

$h_{-2} \times 1/\nu^2$	$(h_{-2} = 10^{10})$	predominant for $\nu < \nu_1 = 1$ Hz.
$h_{-1} \times 1/\nu$	$(h_{-1} = 10^5)$	predominant for $\nu > \nu_1 = 1$ Hz.
h_0	$(h_0 = 1)$	predominant for $\nu > \nu_2 = 10^5$ Hz.
$h_1 \times \nu$	$(h_1 = 10^{-5})$	predominant for $\nu > \nu_3 = 10^{10}$ Hz.
$h_2 \times \nu^2$	$(h_2 = 10^{-10})$	predominant for $\nu < \nu_4 = 10^{15}$ Hz.

The results that should be given by the Allan variance, the Hadamard variance (with $2M = 6$) and the three-samples variance applied to this ideal model of frequency noise are shown in Fig. 6.14 for $T = \tau$ (no dead time).

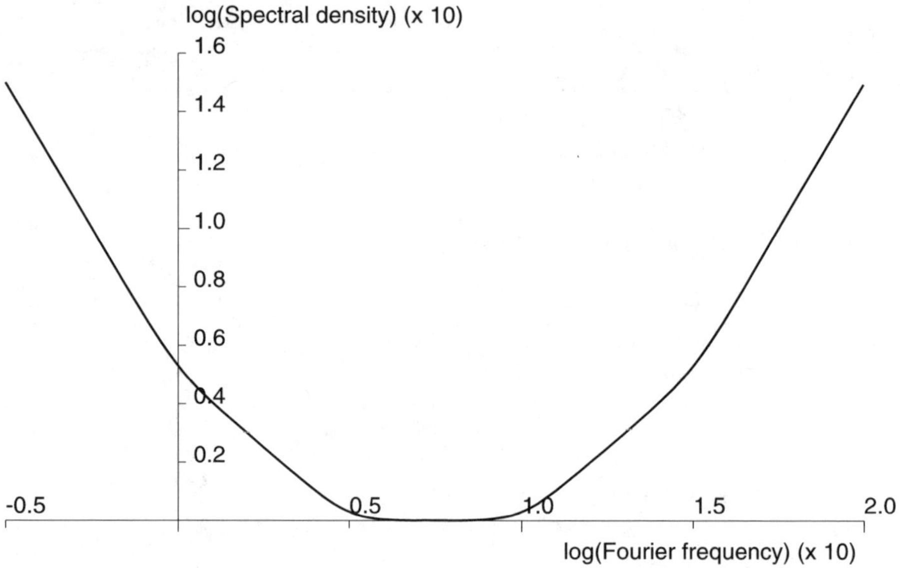

Fig. 6.13. Model of frequency noise showing the five power laws

- The positive slope of the curves (slope +1 in logarithmic coordinates) occurs when the filter associated with the variance (Allan, Hadamard or three-samples) is centered in a Fourier frequency domain where the component $h_{-2} \times 1/\nu^2$ of the noise is predominant.
- The horizontal part of the curves (slope 0 in logarithmic coordinates) occurs when the filter is centered in a Fourier frequency domain where the component $h_{-1} \times 1/\nu$ of the noise is predominant.
- The slope -1 in logarithmic coordinates occurs when the filter is centered in a Fourier frequency domain where the component h_0 is predominant.
- The slope -2 in logarithmic coordinates occurs when the filter is centered in a Fourier frequency domain where the component h_{-1} or h_{-2} is predominant.

The filter associated with every variance is centered at the Fourier frequency $\nu_\mathrm{c} = \nu_\mathrm{s}/2 = 1/2T$.

Consequently:

- The component $h_{-2} \times 1/\nu^2$ of the noise should appear when $T > 1/2\nu_1$.
- The component $h_{-1} \times 1/|\nu|$ of the noise should appear when $1/2\nu_2 < T < 1/2\nu_1$.
- The component $h_0 \times 1/|\nu|$ of the noise should appear when $1/2\nu_3 < T < 1/2\nu_2$.
- The component $h_1 \times 1/|\nu|$ of the noise should appear when $1/2\nu_4 < T < 1/2\nu_3$.
- The component $h_2 \times 1/|\nu|$ of the noise should appear when $T < 1/2\nu_4$.

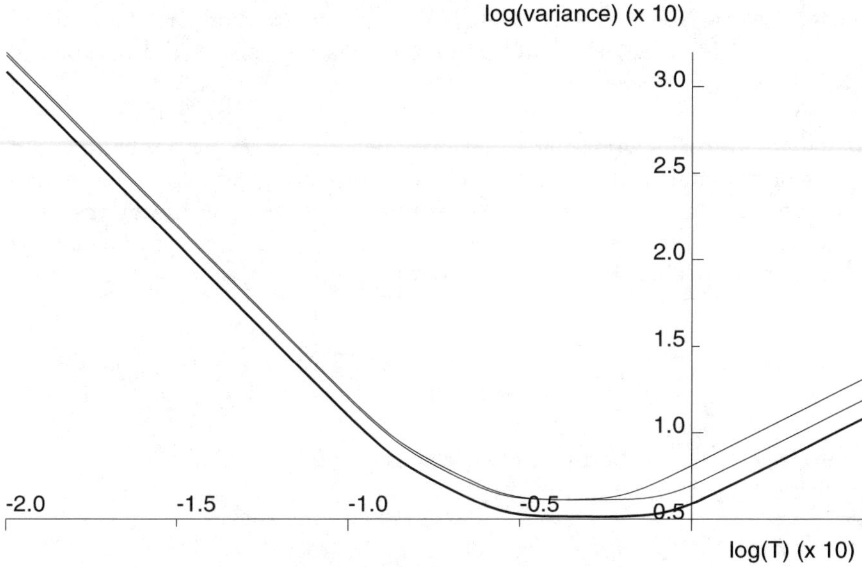

Fig. 6.14. Comparison of Allan (*lower curve*), Hadamard and three-samples (*upper curve*) variances

In fact, the values of the limits indicated above can be strongly modified due to the difference between the sensibility of the variance to each component of the noise. Further, some components can be completely masked by the others.

Consequently, it is important to notice that a given variance can show an erroneous component of the frequency noise for a given sampling time. This is discussed below.

Sensitivity to the Different Noise Components

As a general rule, it is clear that a large difference between the sensitivity of a given variance to the different components of the frequency noise may lead to the masking of the component with the lowest sensitivity by a neighbor component with a high sensitivity.

The Components $h_1 |\nu|$ *and* $h_2\nu^2$

For the three variances, these two components give very similar results (see Table 6.3, Sect. 6.3). Consequently, it is generally not possible to decide from experimental results which one of these two contributions exist in the frequency noise. Figure 6.15 compares two calculated Allan variances when only one of these two frequency noise components exists but not the other. On the other hand, the contribution of the white frequency noise is the same for the two curves. It is clear from this figure that the two curves are very close to each other and cannot be distinguished in experimental results.

If the two components exist, the second one ($h_2\nu^2$) generally masks the first one because the sampling time for which the effect of the component $h_2\nu^2$ becomes smaller than that of the component $h_1|\nu|$ occurs for values where both are masked by some other component. If ν_2 is the Fourier frequency where the two branches corresponding to these two components of the frequency noise intersect, the value of ν_2 is given by

$$h_2\nu_2^2 = h_1\nu_2 \ ,$$
$$\nu_2 = \frac{h_1}{h_2} \ . \tag{6.276}$$

The two corresponding branches of the variance intersect for the value $T_2 = \tau_2$ of the sampling time defined by

$$h_1 g_1 \ln(\pi\nu_{\max}T_2 + k) = h_2 g_2 \nu_{\max} \ ,$$
$$T_2 \approx \frac{1}{\pi\nu_{\max}} \times \exp\left(\frac{g_2}{g_1}\frac{\nu_{\max}}{\nu_2}\right) \ . \tag{6.277}$$

g_1 and g_2 are numerical coefficients that depend on the variance. They are summarized in Table 6.3.

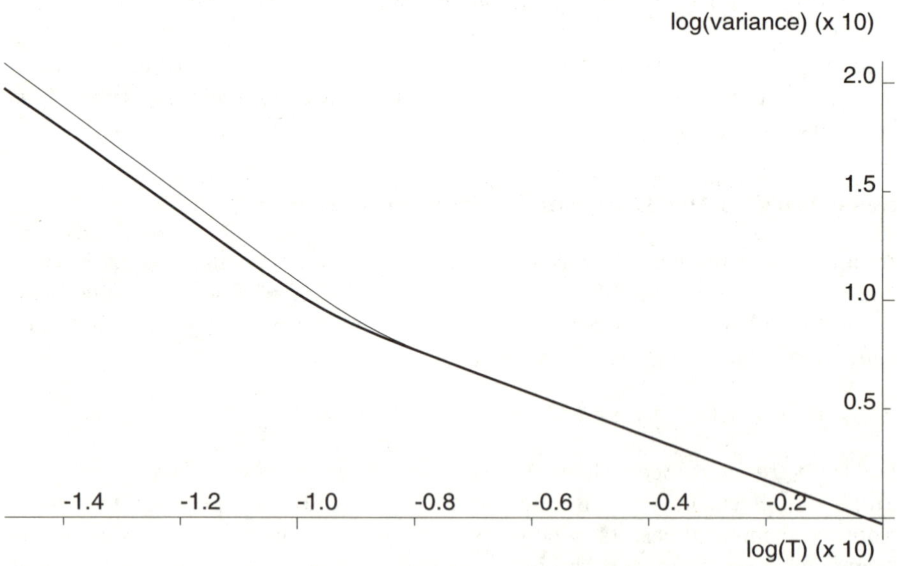

Fig. 6.15. Allan variance with $h_1|\nu|$ noise (*lower curve*) and with $h_2\nu^2$ noise (*upper curve*)

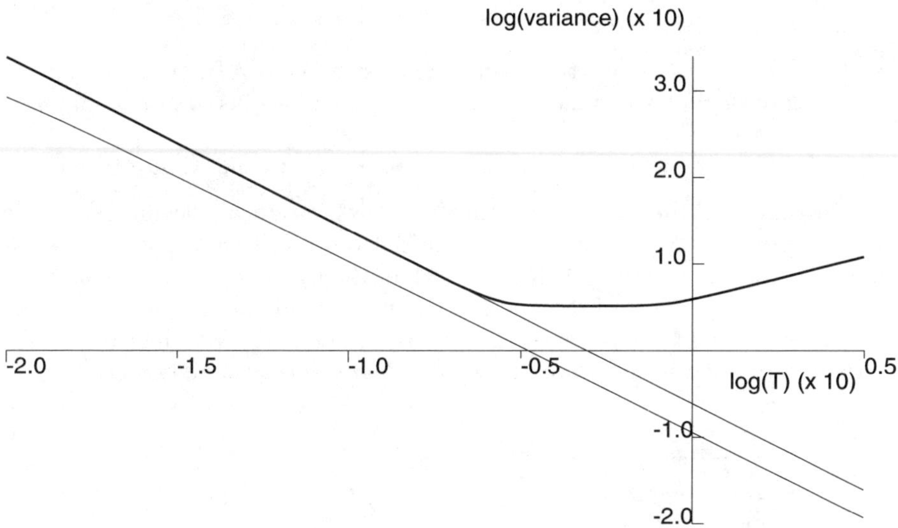

Fig. 6.16. Allan variance and the two asymptotes corresponding to h_1 and h_2

Putting

$$\frac{\nu_{\max}}{\nu_2} = K \gg 1 \, ,$$

we have

$$T_2 \approx \frac{1}{K\pi\nu_2} \times \exp\left(\frac{g_2}{g_1}K\right) \gg \frac{1}{\nu_2} \, . \qquad (6.278)$$

In conclusion, for each variance, the intersection of the two branches is strongly shifted toward the high values of T, where other components are dominant. This is shown in Fig. 6.16 where the Allan variance is calculated in a case where both components $h_1 |\nu|$ and $h_2\nu^2$ of frequency noise exist. The two straight lines are the asymptotes corresponding to the two components. The figure shows that the second component totally masks the first one.

The Components $h_{-2}\nu^{-2}$, $h_{-1}\left|\nu^{-1}\right|$ *and* h_0

The effect of these components can be written for all the variances as $g_i \times T^i$ with $i = 1, 0, -1$, respectively. It is now interesting to study the ratio between the coefficients g_i of two successive branches of a given variance. These branches appear as the sampling time T (assumed equal to τ) is varied. If one of these coefficients is much larger than the coefficient of a following branch, the branch with the larger coefficient will mask the branch with the smaller coefficient. These ratios are shown in Fig. 6.17. They have been calculated with $2M = 6$ for the Hadamard variance.

These values show that

- the Allan variance has the overall smallest difference from g_{-2} to g_0;
- the three-samples variance shows a large difference between g_{-2} and g_{-1}: g_{-2} is 20 dB above g_{-1}; and
- the Hadamard variance ($2M = 6$) is quite close to the Allan variance.

For instance, in the case of the three-samples variance, the limit between the T^1 and T^0 parts of the curve is shifted toward the low values of T, as shown in Fig. 6.14. In Fig. 6.18, the width of the frequency domain where the component $h_{-1} \times |\nu|^{-1}$ of the frequency noise is dominant has been modified, the other ones being unchanged. This noise component is not detected by the three-samples variance, *although it is dominant during three decades.*

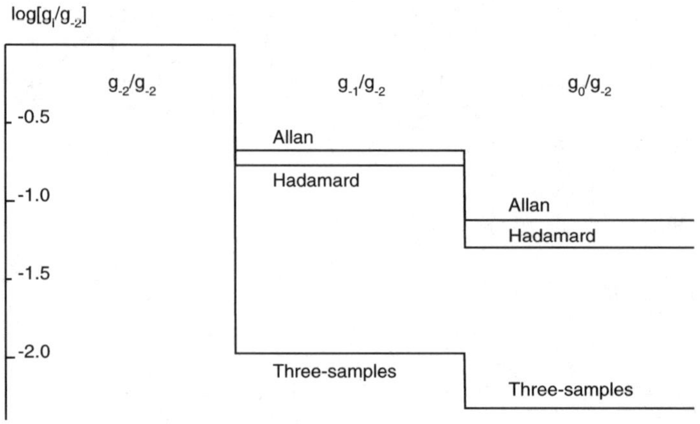

Fig. 6.17. Ratio of the coefficients of the three branches for three variances

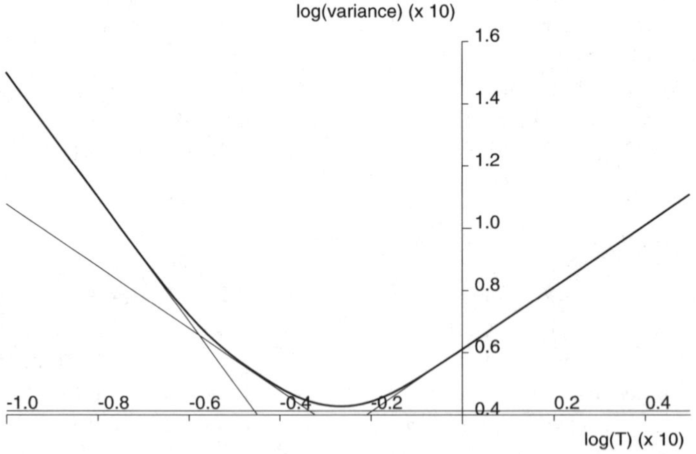

Fig. 6.18. Three-samples variance. The width of the frequency domain where the component $h_{-1} \times |\nu|^{-1}$ of the frequency noise is dominant is three decades

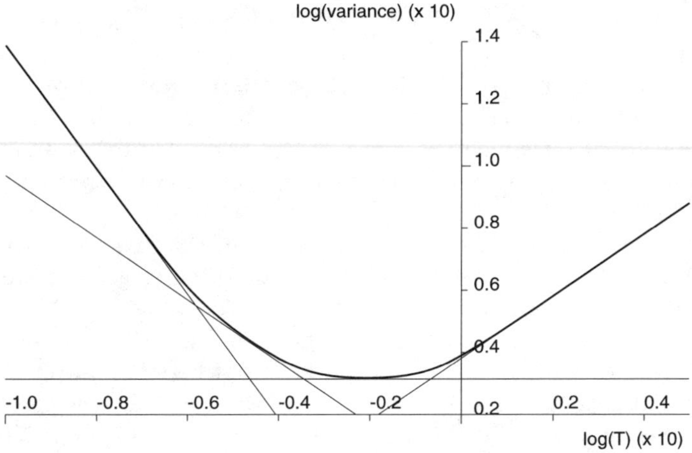

Fig. 6.19. Allan variance. The width of the frequency domain where the component $h_{-1} \times |\nu|^{-1}$ of the frequency noise is dominant is three decades

On the other hand, this component is clearly detected by the Allan variance, (as shown in Fig. 6.19), and by the Hadamard variance.

Linear Frequency Drift

Only the three-samples variance is completely insensitive to any *linear* frequency drift. Allan variance and Hadamard variance are sensitive to a linear frequency drift. Their responses to such a drift are given by (6.201), p. 100 for Allan variance and by (6.256), p. 109 for Hadamard variance. Table 6.4 gives the sensitivity of the three variances to a liner frequency drift characterized by

$$f_{\mathrm{d}}(t) = k \times t \,, \tag{6.279}$$

$$y_{\mathrm{d}}(t) = \frac{f_{\mathrm{d}}(t)}{f_0} = k_y \times t \,. \tag{6.280}$$

The dimension of the coefficient k is Hz^2 (or s^{-2}), that of the coefficient k_y is Hz (or s^{-1}).

The response of the Allan variance and the Hadamard variance to this constant frequency drift is proportional to the square of the sampling time,

Table 6.4. Sensitivity to a constant frequency drift

Variance	Response to a constant frequency drift
Allan	$N/2(N-1) \times (k_y\tau)^2$
Hadamard ($2M = 6$)	$36N/(N-1) \times (k_y\tau)^2$
Three-samples	0

supposed to be equal to the duration of the individual frequency measurements τ.

In order to characterize the importance of the contribution of the frequency drift on the experimental result, the cut-off sampling time T (supposed equal to τ) for which the effect of the frequency drift equals that of the random walk frequency noise (ν^{-2} power law) is compared for the three variances in Table 6.5.

The larger this value, the smaller the effect of the frequency drift on the measurement. Table 6.5 shows that the Allan variance gives better results than the Hadamard variance.

Figures 6.20 and 6.21 show the results given by the Allan variance and the Hadamard variance, respectively, in the case where the output frequency of the oscillator under test is perturbed by the frequency noise defined in Fig. 6.13 and by a constant frequency drift characterized by $k_y = 10^3 \times h_{-2}$. They show that while the component $h_{-2}\nu^{-2}$ (random walk frequency noise) of the frequency noise is quite visible in the case of the Allan variance, it is much less clearly detected in the case of the Hadamard variance, because it is masked by the effect of the constant drift. Of course, the three-samples variance is free of any effect related to a constant frequency drift and should be used when it is important to get rid of a constant frequency drift.

Table 6.5. Cut-off sampling time

Variance	Cut-off sampling time
Allan	$(N-1)/N \times 4\pi^2/3 \times h_{-2}/k_y^2 \approx 13.2 \times h_{-2}/k_y^2$
Hadamard ($2M = 6$)	$(N-1)/N \times 2\pi^2/9 \times h_{-2}/k_y^2 \approx 2.19 \times h_{-2}/k_y^2$
Three-samples	No effect

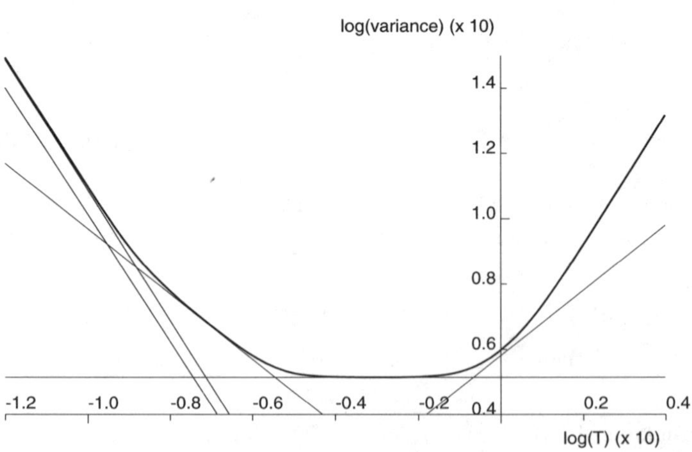

Fig. 6.20. Allan variance with a constant frequency drift

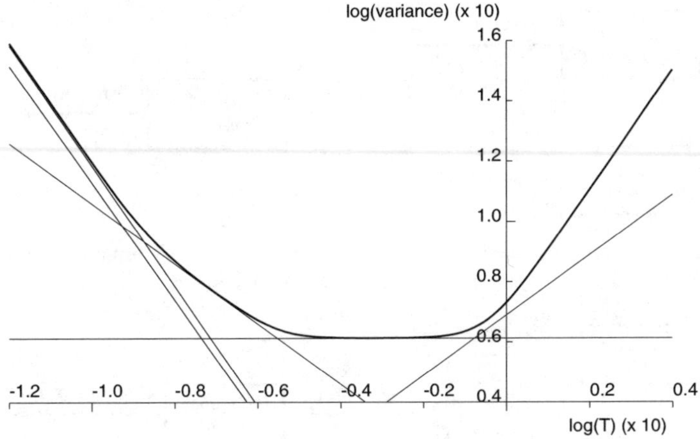

Fig. 6.21. Hadamard variance with a constant frequency drift

6.2.8 Practical Time Domain Measurements of Frequency Stability

Variance and Standard Deviation

The stability of an ultra-stable oscillator, when measured in the time domain, is usually characterized by the (Allan or Hadamard or any other ...) standard deviation.

The standard deviation is the square root of the variance: the Allan standard deviation, for instance, is the square root of the Allan variance.

Frequency Measurements

The first step of the time domain measurements of the frequency stability is a series of frequency measurements. The duration of each of these measurements is τ. The following two situations are possible.

1. There exists an oscillator whose stability is better than that of the oscillator under test. This is the reference oscillator.
   In this case, it is easy to measure the mean value of the frequency of the oscillator under testing during successive time intervals of duration τ by frequency comparison with this reference oscillator. For instance, the reference oscillator can be used as a local oscillator to down convert the frequency of the oscillator under test. The down converted frequency is such that it can be easily measured by a frequency counter (see Fig. 6.22).
2. The oscillator under testing is supposed to have the best stability properties and there is consequently no other oscillator that can be used like a reference oscillator.

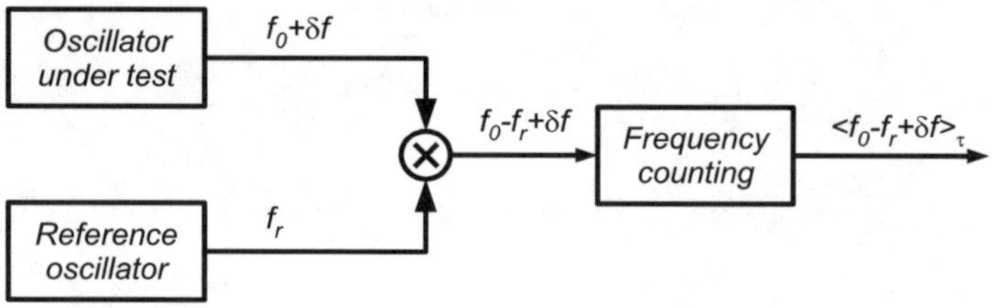

Fig. 6.22. Frequency measurement in the case where a reference oscillator can be used

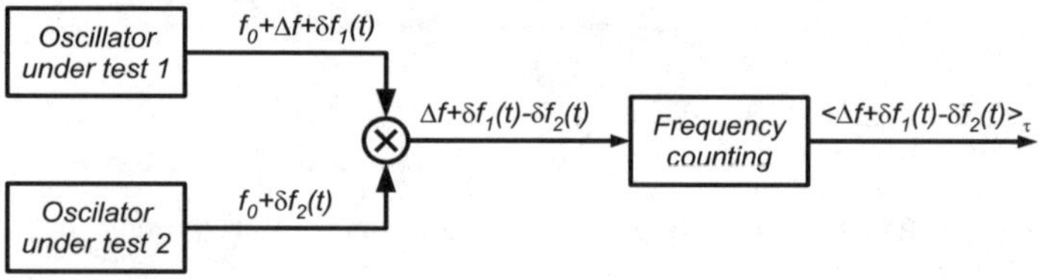

Fig. 6.23. Stability measurement with two identical oscillators

In this case, the frequency of two specimens of the same oscillator are compared and the fluctuations of the frequency difference between the two specimens are studied. It is possible to shift the frequency of one of the two oscillators in order to obtain a difference frequency that can be easily measured (see Fig. 6.23).
It is assumed that the two specimens have the same stability properties and that the frequency fluctuations are not correlated. Consequently, the mean square value of the fluctuations of the difference of the frequency of the two oscillators equals twice the mean square value of the fluctuations of one of the oscillators.
The variance of each oscillators is the measured variance of the difference of the frequencies of the two oscillators divided by two, and the deviation of each oscillator is the measured deviation divided by $\sqrt{2}$.

6.2.9 Typical Values

A few stability values of some frequency standards are given below.

A Microfabricated Atomic Clock

This device is described in [72]. It is a passive atomic frequency standard that uses the hyperfine transition of cesium as a reference (see also [50]).

The main parameters of its physical package are the following:

- volume: $9.5\,\text{mm}^3$, and
- power dissipation: 75 mW.

It is clear that the main objective of this kind of device is not to obtain the best values of stability and accuracy but to enable low-cost mass production.

Consequently, it is possible to characterize its frequency stability with a reference oscillator that is much more stable.

It shows a linear fractional frequency drift that has been measured by direct comparison with a stable oscillator. This drift is about -2×10^{-8}/day [72], which corresponds to

$$d_y = 2.3 \times 10^{-13}\,\mathrm{s}^{-1}\,. \tag{6.281}$$

This drift should give a component $\sigma'_y(\tau)$ to the Allan deviation equal to

$$\begin{aligned}\sigma'_t(\tau) &= \sqrt{\frac{1}{2}} d_y \tau \\ &= 1.6 \times 10^{-13} \times \tau\,.\end{aligned} \tag{6.282}$$

This is in good agreement with the experimental Allan deviation shown in Fig. 3 of [72].

For small values of the sampling time, the experimental Allan deviation is fitted by $2.4 \times 10^{-10}/\sqrt{\tau}$, which corresponds to a white frequency noise with a spectral density h_0 equal to

$$h_0 = 1.15 \times 10^{-19}\,\mathrm{Hz}^{-1}\,. \tag{6.283}$$

Because the frequency drift is measured directly, it is possible to remove it from the calculation of the variance. This has been done in [72] and the long-term Allan deviation with this drift removed is approximately

$$\sigma_y(\tau) = 1.5 \times 10^{-12} \times \tau^{1/2}\,. \tag{6.284}$$

This corresponds to a component of the frequency noise $h_{-2}\nu^{-2}$ (random walk frequency noise) with

$$h_{-2} = 3.4 \times 10^{-25}\,\mathrm{Hz}\,. \tag{6.285}$$

The flicker frequency noise component, if any, is masked by these two components: no horizontal floor is visible on the Allan deviation graph of this device to the level of 2.5×10^{-11} for a sampling time of 250 s. Consequently, the dimensionless coefficient h_{-1} of this component is smaller than

$$\begin{aligned}h_{-1,\max} &= \frac{\left(2.5 \times 10^{-11}\right)^2}{1.39} \\ &= 4.5 \times 10^{-22}\,.\end{aligned} \tag{6.286}$$

Figure 6.24 shows the spectral density of the frequency noise of this oscillator deduced from the Allan variance measurement.

The measurements reported in [72] were taken for sampling times between 1 and about 2000 s. The center frequency of the filter equivalent to the Allan algorithm was varied between 0.5 and 2.5×10^{-4} Hz. The curve of Fig. 6.24 is consequently limited to this range of frequencies.

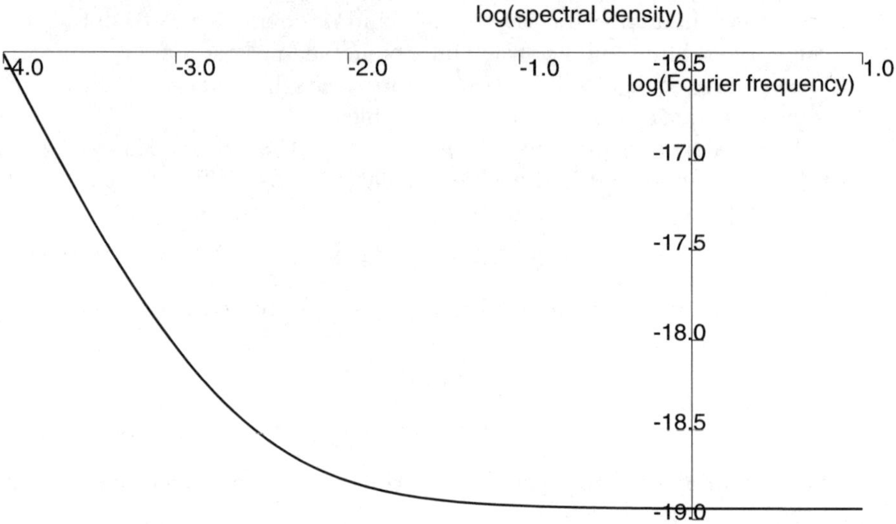

Fig. 6.24. Frequency noise spectral density of the oscillator of [72]

Commercially Available Active Hydrogen Maser

The hydrogen maser is a device that provides excellent medium-term and long-term stability.

The results of Table 6.6 are those published by Precise Time and Frequency, Inc. (PTF) for the CHI-75 active hydrogen maser [104] (no dead time: the sampling time equals the frequency measurement duration). They give the Allan deviation for some values of the sampling time. They are "typical values" and do not describe a given device.

These results show two components of the frequency noise.

1. The first component is dominant for the sampling times larger than about 100 s. The corresponding Allan deviation is proportional to $1/\sqrt{T}$, i.e.

Table 6.6. Typical values of the Allan standard deviation of the CHI-75 active hydrogen maser

Sampling time (s)	Allan deviation
1	2×10^{-13}
10	2.5×10^{-14}
100	5×10^{-15}
1000	1.6×10^{-15}
3600	1×10^{-15}
10000	7×10^{-16}
100000	2×10^{-16}

Allan variance is proportional to $1/T$. It is consequently a white frequency noise (see Table 6.3).

2. The second component is dominant for the sampling times $1\,\mathrm{s} \leq T \leq 100\,\mathrm{s}$ and leads to an Allan deviation proportional to $1/T$, i.e. an Allan variance proportional to $1/T^2$. It is one of the two frequency noise components flicker phase noise ($h_1\,|\nu|$) or white phase noise ($h_2\nu^2$), as shown in Fig. 6.15.

Commercially Available Cesium Frequency Standard

The results of Table 6.7 are those published by Symmetricom for the 5071A Primary Frequency Standard (High Performance) [113]. They give the Allan deviation for some values of the sampling time. Moreover, the guaranteed flicker floor is 1×10^{-14}.

These values show that the dominant component of the frequency noise all over the range $0.01 \rightarrow 10^6$ s is the white frequency noise.

Table 6.7. Specification values of the Allan deviation for the 5071A primary frequency standard (high performance)

Sampling time (s)	Allan deviation
0.01	7.5×10^{-11}
1	5×10^{-12}
100	8.5×10^{-13}
10 000	8.5×10^{-14}
2.592×10^6	1×10^{-14}

Atomic Fountain Primary Frequency Standard

These frequency sources are designed in order to obtain the best values of accuracy and long-term stability. They are laboratory devices that must be handled by highly specialized staff. They are described, for instance, in [125, 57, 101, 14, 124].

The stability values given in [124] have been obtained using the following protocol.

1. Since the stability of these devices are among the best values, only a comparison of two identical apparatus can provide a reliable result.
2. Two identical atomic fountains have consequently been built. The microwave signal probing the atomic transition of each apparatus (see Sect. 2.3.5) is synthesized from a cryogenic sapphire oscillator (CSO) [92] phase-locked to a hydrogen maser with a long time constant (i.e. a low

> cut-off frequency of the low-pass filter of the loop). The short-term stability ($\approx 10^{-15}$) is that of the CSO up to 800 s, whereas the long-term stability is given by the maser. Each frequency chain uses a 7.3 MHz signal generated by a computer-controlled high-resolution synthesizer. The detected atomic transition probability determines the frequency corrections applied to the microwave probing signal via the synthesizer. These corrections are the base for the accuracy and frequency stability evaluation of each fountain.

The Allan deviation of the frequency difference between the two frequency standards is $5 \times 10^{-14}/\sqrt{T}$ for sampling times T between 100 s and 3×10^4 s (the value of the standard deviation for $T = 3 \times 10^4$ s is 3×10^{-16}). This result corresponds to a white frequency noise whose fractional spectral density is $p_y(\nu) = 5 \times 10^{-27}\,\mathrm{Hz}^{-1}$.

Part III

Applications

7 Time and Frequency Metrology

The second is one of the seven SI base units. Its practical realization provides the best standard uncertainty. The uncertainties for the seven SI base units are shown in Table 7.1 [105].

Furthermore, the definition of the unit of length (meter) relies on the realization of the second: "*The meter is the length of the path traveled by light in vacuum during a time interval of* 1/299 792 458 *of a second.*" [19, 20].

Special care must therefore be taken in the process of dissemination of this SI unit to the users. The needs of the final users may be the following:

1. Time or frequency.
   - In the first case, the user needs to measure the value of time intervals and/or to locate a given event on the time axis, i.e. to measure the time interval between this event and a conventional origin; this is the date of the event. The user needs the unit of time and the conventional origin and a device that counts the number of units of times. This is a clock.
   - In the second case, there is no need for any conventional origin on the time axis. The user wants to measure the frequency of a periodic phenomenon. For this he needs the unit of time and a device that counts the number of periods of the phenomenon during that unit of time.
2. Accuracy and/or stability and/or another metrological property and/or particular environment, etc.
   According to the application, the requirements on any of these qualities may be more or less important.

Table 7.1. The uncertainties of the best realizations of the seven SI base units

SI base unit	Symbol	Uncertainty
Second	s	1×10^{-15}
Meter	m	1×10^{-12}
Kilogram	kg	2×10^{-9}
Candela	cd	1×10^{-4}
Mole	mol	2×10^{-9}
Ampere	A	9×10^{-8}
Kelvin	K	3×10^{-7}

To meet these needs, a great variety of devices has been built, each one being adapted to a special use. These devices must be characterized and compared in order to

1. define a time scale that allows one to date the events and realize the unit of time, and
2. define some reference frequencies that realize the unit of frequency.

7.1 Time Scales

A time scale is a virtual clock from which any user should be able to measure the value of the parameter time corresponding to any event. An ideal time scale should meet the following requirements:

1. It should be accessible to any user.
2. It should rely on the SI time unit.
3. It should permit the measurement of the time interval between two events, according to the SI time unit definition.
4. It should permit the measurement of the time interval between a given event and a conventional origin. This measurement provides the date of the given event in that time scale.
5. It should take relativity [44] into account. In the framework of general relativity, the physical world is described in four-dimensional space–time reference systems. The four dimensions of a specified reference system are the three spatial coordinates and $x_0 = c \times t$, where c is the velocity of light in vacuum and t is the coordinate time in this reference system. Coordinate time provides an unambiguous way of dating events *in a specified reference system.* This is the time coordinate to be used in the theory of motion *referred to this system.*
   In a general way, the observed time of an observer described in a specified reference system differs from the coordinate time of this reference system. The relation of the observed time to the coordinate time of the reference system depends on the velocity of the observer in the reference system, mass repartition and energy.
   The observed time of an observer is called the proper time of this observer.
6. It should satisfy some particular constraints imposed by some particular applications. For instance, the time scale regulating ordinary human activities should take into account the rotation of the Earth.

One single time scale cannot meet all of these requirements in every context.

1. A time scale defined in a specified reference system is not suited for applications in another reference system.
2. A time scale defined as the coordinate time of a specified system may not be suited for applications involving a proper time.

3. Many human activities are dependent on the length of the day and involve a time scale linked to this length. Since the Earth's rotation is not uniform, and consequently the length of the day is not constant, a time scale defined on a constant second basis, such as the one produced by an atomic clock, would not fit the needs of many human activities.

The reference time scale is TAI (temps atomique international, international atomic time), which is built on the SI unit produced by atomic clocks. The other time scales are connected to TAI and meet the needs of different kinds of users.

The main time scales are briefly described in the following.

7.1.1 The Main Time Scales in Use

TAI

The first definition of TAI was that submitted by CCDS[1] in 1970 to CIPM [33] and [67][2]: "*International Atomic Time (TAI) is the time reference coordinate established by the Bureau International de l'Heure on the basis of the readings of atomic clocks operating in various establishments in accordance with the definition of the second, the unit of time of the International System of Units.*"

This definition did not specify the reference system to which it applies. It was completed in 1980 [32] and [100][3]: "*TAI is a coordinate time scale defined in a geocentric reference frame with the SI second as realized on the rotating geoid* [4] *as the scale unit.*"

The clocks included in TAI must therefore be placed at sea level or a correction must be made to take the actual altitude (the effect of terrestrial mass) and the latitude (the effect of the centripetal potential due to the Earth's rotation) into account [102].

The gravitational effect within the theory of relativity affects the frequency of any clock: a clock in a smaller gravitational field is faster than a clock in a higher gravitation. In the vicinity of the Earth's surface, the relative effect is about 1×10^{-16} for 1 meter of altitude difference; a clock

[1] CCDS – Comité Consultatif de la Seconde – has been renamed as CCTF – Comité Consultatif du Temps et des Fréquences, Consultative Committee for Time and Frequency – by CIPM in 1997.

[2] The official text is in French and reads: "Le Temps atomique international est la coordonnée de repérage temporel établie par le Bureau international de l'heure sur la base des indications d'horloges atomiques fonctionnant dans divers établissements conformément à la définition de la seconde, unité de temps du Système international d'unités."

[3] The official French text reads: "Le TAI est une échelle de temps-coordonnée définie dans un repère de référence géocentrique avec comme unité d'échelle la seconde du SI telle qu'elle est réalisée sur le géoïde en rotation."

[4] The geoid is the gravity equipotential surface (surface of fixed potential value) that coincides on average with the mean sea level.

located at a height of 1 646 m (NIST, Boulder, CO, for instance) will be approximately 140 nanoseconds faster per day than a clock located at a height of 60 m (LNE-SYRTE, Paris, for instance).

This definition was completed by the International Astronomical Union in 1991, Resolution A4: "*TAI is a realized time scale whose ideal form, neglecting a constant offset of 32.184 s, is terrestrial time (TT), itself related to the time coordinate of the geocentric reference frame, geocentric coordinate time (TGC), by a constant rate.*"

Responsibility for TAI was accepted by the CIPM from the Bureau International de l'Heure on 1 January 1988.

TAI consists of two steps:

1. A weighted average based on a great number of clocks maintained under metrological conditions in about fifty laboratories[5] is calculated. The algorithm used is optimized for long-term stability, which requires the observation of the behavior of clocks over long periods. In particular, the relative weight is allotted to each clock after examination of its behavior over the duration of the period considered. In consequence, TAI is a deferred-time time scale, available with a delay of a few days[6].
   The time scale resulting from this computation is called EAL (échelle atomique libre – free atomic scale). The medium-term stability of this intermediate scale (Allan deviation for averaging times of 20 to 40 days) is estimated to be 0.6×10^{-15} over the period January 1999 to June 2004 [21].
2. The frequency error of this intermediate scale is evaluated by comparing it with some primary frequency standards (PFS). This, of course, requires the application of a correction to connect the proper time of each of these frequency standard to the coordinate time of the rotating geoid.
   In January 2006, for instance, the relative difference between the frequency of TAL and the frequency of the ensemble of primary frequency standards was measured to be 6.815×10^{-13}, with an uncertainty of 0.009×10^{-13} [24].
   The primary frequency standards implied in this evaluation are shown in Table 7.2.
   A weight is allotted to these primary frequency standards, according to the uncertainty affecting their error budget. The result of these comparisons allow TAI to be corrected in such a way that
   - its stability is the stability of EAL, and
   - its accuracy is the accuracy of the ensemble of the primary frequency standards.

---

[5] For instance, in the note describing the relative weights of the clocks participating in the computation of TAI published by the BIPM on 9 February 2006 [25], 321 clocks from 47 institutes are listed.

[6] For instance, the circular T number 217 published by the BIPM on 9 February 2006 [22] gives the information for the period 30 December 2005 to 29 January 2006.

Table 7.2. Primary frequency standards involved in the accuracy of TAI

Acronym	Full name	Location	PFS
NIST	National Institute of Standards and Technology	Boulder, CO., USA	F1
PTB	Physikalisch-Technische Bundesanstalt	Braunschweig, Germany	CS1 CS2
OP	Observatoire de Paris	Paris, France	FO2 JPO
IEN	Istituto Elettrotecnico Nazionale Galileo Ferraris	Turin, Italy	CSF
NICT	National Institute of Information and Communications Technology	Japan	NICT01
NMIJ	National Metrology Institute of Japan	Tsukuba, Japan	F1

To obtain this result, the frequency correction is of the order of the frequency instability and is applied at intervals of two months. Circular T, edited by BIPM, gives the correction applied to EAL. For instance, circular T number 217 [22] gives the following information:

– The relative steering correction applied to EAL during the period 53 734–53 764 (dates are given in MJD, modified Julian day [137] and correspond to the interval 30 December 2005 to 29 January 2006) was

$$\left(f_{\mathrm{EAL}} - f_{\mathrm{TAI}}\right)/f_{\mathrm{TAI}} = 6.850 \times 10^{-13} \,.$$

– The relative steering correction applied to EAL during the period 53 764–53 794 (29 January 2006 to 28 February 2006) was

$$\left(f_{\mathrm{EAL}} - f_{\mathrm{TAI}}\right)/f_{\mathrm{TAI}} = 6.844 \times 10^{-13} \,.$$

Consequently, a correction of 0.6×10^{-15} was applied to TAI on 29 January 2006.

Time Scales for Astronomy

Two different groups of time scales are used in astronomy.

1. Time scales based on the SI second. These time scales rely on atomic phenomenon and are related to TAI.
2. Time scales based on the rotation of the Earth. It is known that the length of the day is not constant and has unpredictable variable components. It must consequently be continuously monitored through astronomical observations, which is done primarily with very long baseline interferometry (VLBI, see Sect. 9).

Before the introduction of atomic clocks, all time scales were tied to the rotation of the Earth.

Time Scales Based on the Rotation of the Earth

Time scales that are based on the rotation of the Earth are used in astronomical applications, such as telescope pointing, which depend on the geographic location of the observer. The history of these time scales is briefly described in the Introduction.

1. Sidereal time and local sidereal time are the hour angle of the equinox with respect to the Greenwich meridian or the local meridian, respectively.
2. Universal time (UT). There are two forms of universal time: UT0 and UT1, the last one being the form that is almost always used. The first definitions of UT rely on sidereal time: The observed mean sidereal time at Greenwich (measured from the position of some reference stars) is called GMST; it is transformed to give a mean solar time, the first universal time standard, UT0. This in turn gives rise to a time standard, UT1, that corrects for nutation and pole movement (two further periodic terms are removed from UT1 to form UT2). Some effects due to tidal forces remain, which slow the Earth's rotation. In 2000, IAU[7] redefined UT1 to be a linear function of the Earth's rotation angle, θ, which is the geocentric angle between two directions in the equatorial plane called, respectively, the celestial intermediate origin (CIO) and the terrestrial intermediate origin (TIO) [65][8]. TIO rotates with the Earth, while CIO has no instantaneous rotation around the Earth's axis, so that θ is a direct measure of the Earth's rotational motion that is unaffected by precession and nutation.

All these time scales are affected by the variations in the Earth's rate of rotation, which are unpredictable and must be measured through astronomical observations. The lengths of the sidereal and UT1 seconds are, therefore, not precisely constant when expressed in a uniform time scale based on the SI second. The accumulated difference in time measured by a clock keeping SI seconds on the geoid from that measured by the rotation of the Earth is $\Delta T = \mathrm{TT} - \mathrm{UT1}$. The long-term trend for ΔT is to increase because of the tidal deceleration of the Earth's rotation.

Times Scales Based on the SI Second

Since the SI second results from the observation of an atomic phenomenon, SI-based time scales can be constructed or hypothesized anywhere in the universe, which is not the case of the time scales based on the Earth's rotation.

[7] IAU: The International Astronomical Union.

[8] These origins were called the celestial ephemeris origin and the terrestrial ephemeris origin in [65].

Successive IAU recommendations have been emitted to define time scales for two different reference systems. Each time scale, corresponding to a given reference system, is the coordinate to be used in the theory of motions referred to this system.

These time scales are based on the SI second and are the time coordinates of the two following reference systems:

- GCRS is geocentric, which means that the origin of the spatial coordinates is located at the barycenter of the Earth. This is used to describe the motions relative to the Earth.
- BCRS is centered at the barycenter of the solar system.

Both reference systems are non-rotating with respect to the ensemble of distant extragalactic objects. For each of these two reference systems, the time coordinates is the time scale, which should be delivered by an ideal TAI clock, which is

- located on the origin of the spatial coordinates (i.e. on the barycenter of the Earth) in the case of the geocentric reference system,
- located far from any source of gravitational field in the case of the barycentric reference system.

The origin of the time coordinates of both reference systems is chosen in such a way that at the instant 1977 January $01^{\mathrm{d}}00^{\mathrm{h}}00^{\mathrm{m}}00^{\mathrm{s}}$, the value of the time coordinate is 1977 January $01^{\mathrm{d}}00^{\mathrm{h}}00^{\mathrm{m}}32.184^{\mathrm{s}}$.

These two time scales are called

- the geocentric coordinate time (TCG: Temps-coordonnée géocentrique in French) for GCRS, and
- the barycentric coordinate time (TCB: temps-coordonnée barycentrique in French) for the barycentric celestial reference system (BCRS).

No real physical clock can deliver these two time scales, but they are related to TAI and to a third astronomical time scale that is widely used for geocentric and topocentric ephemeris:

- the terrestrial time (TT), which has the same origin as TCG and TCB, but whose unit is the SI second. Its rate is consequently exactly equal to that of TAI and the only difference between these two scales is the offset of 32.184 s.

  TT is conventionally realized by $\mathrm{TT} = \mathrm{TAI} + 32.184\,\mathrm{s}$.

In fact, TCG coincides with TT except that it omits the gravitational effect of the Earth itself (and the time dilation caused by its rotation), since the virtual clock that delivers it is situated at the barycenter of the Earth.

Consequently, the TCG is faster than TAI and TT by a factor of $6.969290134 \times 10^{-10}$. This factor is now considered as a defining constant and is not subject to further revision.

TCG differs from TCB because the first one is the SI time scale of an observer submitted to the gravitation of the Sun and of the other planets and in motion around the Sun.

Coordinated Universal Time (UTC)

The time scale TAI, based on the transition of atoms, is now the best approximation of the ideal time scale with the best regularity possible. On the other hand, due to the irregularities of the rotation of the Earth and to the fact that everyday activities are based on the length of the day, a perfect regular time scale, which diverges relative to UT, is not suited for these activities.

For this reason, UTC is a time scale based on the SI second, and possesses the regularity of TAI, but nevertheless never deviates from the time scale UT1 by more than 0.9 s; it is consequently the bridge between atomic and ephemeris times. This result is obtained very simply by inserting leap seconds (which can be positive or negative) in the scale UTC when the difference between UTC and UT1 reaches 0.9 s. In this way

- the difference UTC-UT1 is always smaller than 0.9 s,
- the regularity of the ticks of UTC is the regularity of TAI, but
- events occurring during a leap second must be considered with special care.

The leap seconds that have been inserted in TAI since 1972 are summarized in Table 7.3.

UTC is the time scale in common use (broadcast by radio, television, the telephone, etc.). In addition, the legal time of most countries is offset from UTC by an integer number of hours (time zones and "summer time"). National time-service laboratories maintain an approximation of UTC known as UTC(k) for laboratory k (see Sect. 7.1.1). The differences between UTC(k) and UTC are in general no more than a few hundreds of nanoseconds.

Table 7.3. Dates of the leap seconds since 1972

1972 Jul. 1	1973 Jan. 1	1974 Jan. 1	1975 Jan. 1	1976 Jan. 1
1977 Jan. 1	1978 Jan. 1	1979 Jan. 1	1980 Jan. 1	1981 Jul. 1
1982 Jul. 1	1983 Jul. 1	1985 Jul. 1	1988 Jan. 1	1990 Jan. 1
1991 Jan. 1	1992 Jul. 1	1993 Jul. 1	1994 Jul. 1	1996 Jan. 1
1997 Jul. 1	1999 Jan. 1	2006 Jan. 1		

The difference between UTC and TAI was 33 s on 1 January 2006

The GPS Time Scale

GPS (Global Positioning System) is described elsewhere in this book (see Sect. 8). It relies on precise time measurements and provides a precise time scale called the GPS time scale or GPS time. This time scale is accessible to any GPS receiver.

The ensemble of clocks used to compute the GPS time scale is made up of the clocks of all the operational satellites and of the monitor stations.

Moreover, GPS time is steered to UTC(USNO) from which it does not deviate by more than one microsecond. The exact difference between GPS and UTC(USNO) is contained in the navigation message emitted by each satellite, which gives the difference and rate of GPS relative to UTC(USNO). UTC(USNO) itself is kept very close to the international UTC as maintained by BIPM, and the exact difference between UTC(USNO) and UTC is published monthly in the circular of the Bureau International des Poids et Mesures [22]. The GPS time scale was initialized on 6 January 1980, and was then set equal to UTC (see Sect. 7.1.1). Since this epoch, leap seconds have been added to UTC (see Table 7.3) but no leap second has been added to GPS time; the two scales thus differ by an integer number of seconds; the difference was for instance 14 s on 1 January 2006.

This difference and the difference of the time broadcasted by each satellite and the GPS time are broadcast in the message emitted by the satellites. Any GPS receiver can consequently compute UTC from the received message.

Local Time Scales

Many national measurement institutions (NMI) all over the world have built their own time scale from a set of local clocks. The listing of these 67[9] institutions can be found in [18]. The countries involved are given in alphabetic order in Table 7.4.

7.1.2 Algorithms for the Generation of Time Scales

The algorithms used to construct a time scale from an ensemble of clocks rely on a model of the participating clocks.

Model of the Clocks

The Clocks Involved in Time Scales

The clocks involved in the time scales are mainly commercial cesium frequency standards and hydrogen masers. In the note [25], 321 clocks are mentioned, among them 4 primary frequency standards and 72 hydrogen masers. The others (245) are commercial cesium frequency standards.

Commercial frequency standards are neither more stable nor more accurate. Nevertheless, they are involved in the construction of time scales due to their reliability.

---

[9] in 2006

Table 7.4. Countries maintaining a local approximation of UTC and/or an independent local time scale

Argentina	Australia	Austria
Belgium	Brazil	Bulgary
Canada	Chile	China
Czech Republic	Egypt	France
Germany	Hong Kong	Hungary
India	Israel	Italy
Japan	Lithuania	Malaysia
Mexico	The Netherlands	New Zealand
Norway	Panama	Poland
Portugal	The Republic of Korea	Romania
Russia	Singapore	Slovakia
South Africa	Spain	Sweden
Switzerland	Taiwan	Thailand
Turkey	United Kingdom	USA

The behavior of each clock contributing to a given time scale is described by a limited number of deterministic parameters and by the noise that perturbs its output frequency. One must keep in mind that the stability of commercial cesium frequency standards is much better than their accuracy.

Deterministic Parameters

Each clock i of a given time scale can be characterized at a given time t (this time is given by the time scale) by three parameters as follows:

- The time difference $\delta t_i(t)$ of the clock compared to the time given by the time scale:

$$\delta t_i(t) = t_i(t) - t \ .$$

- The dimensionless reduced frequency difference $\delta y_i(t)$ of the clock compared to the frequency of the time scale:

$$\delta y_i(t) = \frac{f_i(t) - f_0}{f_0} = y_i - 1 \ ,$$

  where f_0 is the value of the time scale frequency.
- The fractional frequency aging d_i of the clock, which produces a deterministic linear variation of its reduced frequency:

$$\Delta y_i(t) = d_i \times (t - t_0) \ ,$$

  where t_0 is an arbitrary time origin and d_i is expressed in s^{-1}.

Knowing the value of these three parameters at a given time t_0, it is easy to calculate their value at any other time $t_0 + \tau$.

The time unit of the time scale is a given number n_0 of the period $T_0 = 1/f_0$ of the virtual clock associated to it; the time unit produced by the clock i is different from the time unit produced by the time scale. Its value $u_i(t)$ is

$$\begin{aligned} u_i(t) &= n_0 \times \frac{1}{f_i(t)} \\ &= n_0 \times \frac{1}{f_0} \times \frac{f_0}{f_i(t)} \\ &= \frac{1}{y_i(t)} \,. \end{aligned} \tag{7.1}$$

If the fractional frequency error at time t_0 is $\delta y_i(t_0)$, its value for any time t is

$$\delta y_i(t) = \delta y_i(t_0) + d_i(t - t_0) \,. \tag{7.2}$$

The unit time of this clock i varies

$$u_i(t) = \frac{1}{y_i(t)} \,. \tag{7.3}$$

Consequently, when a time τ (as measured by the time scale) elapses, the clock i measures (with its wrong time unit)

$$\begin{aligned} \tau + \delta\tau_i &= \int_{t_0}^{t_0+\tau} \frac{dt}{u_i(t)} \\ &= \int_{t_0}^{t_0+\tau} y_i(t)\mathrm{d}t \qquad (7.4) \\ &= \int_{t_0}^{t_0+\tau} \left(y_i(t_0) + d_i(t - t_0)\right) \mathrm{d}t \\ &= \tau \times y_i(t_0) + \int_{t_0}^{t_0+\tau} d_i(t - t_0)\mathrm{d}t \\ &= \tau \times y_i(t_0) + \frac{d_i\tau^2}{2} \qquad (7.5) \\ &= \tau \times [1 + \delta y_i(t_0)] + \frac{d_i\tau^2}{2} \,. \qquad (7.6) \end{aligned}$$

The time error of the clock i consequently increased by $\delta\tau_i$,

$$\delta\tau_i = \delta y_i(t_0)\tau + d_i\tau^2/2 \,. \tag{7.7}$$

The values of the three parameters at time $t_0 + \tau$ are consequently

Parameter	At time t_0	At time $t_0 + \tau$
Fractional frequency error	$y_i(t_0)$	$y_i(t_0) + d^i\tau$
Time error	$\delta t_i(t_0)$	$\delta t_i(t_0) + \delta y_i(t_0)\tau + d_i\tau^2/2$
Frequency aging	d^i	d^i

Random Contributions

The various noise sources that can affect the output frequency of the clock i must also be taken into account. They are described in Sect. 6.2.2; how they are detected by the variances is summarized in Sect. 6.2.7.

The possible noise processes are

- white phase noise:
  spectral density: $h_2\nu^2$,
  Allan variance: $h_2 \times 3\nu_{\max}/4\pi^2 \times \tau^{-2}$,
- flicker phase noise:
  spectral density: $h_1 |\nu|$,
  Allan variance: $h_1 \times 3\left[\ln(\pi\nu_{\max}\tau) + 1.04\right]/4\pi^2 \times \tau^{-2}$,
- white frequency noise:
  spectral density h_0,
  Allan variance: $h_0/2 \times \tau^{-1}$,
- flicker frequency noise:
  spectral density $h_{-1} |\nu|^{-1}$,
  Allan variance: $h_{-1} \times 1.39$,
- random walk frequency noise:
  spectral density $h_{-2}\nu^{-2}$,
  Allan variance: $h_{-2} \times 2\pi^2/3 \times \tau$.

Algorithms

Different algorithms are used to construct the various time scales such as TAI, GPS and the local time scales. All of them rely on the prediction and observation of the time given by the individual clocks involved in the time scale.

Among the N clocks C_j $(1 \leq j \leq N)$ participating in the time scale, one of them, the clock C_{m}, is used as a relay to get access to the time scale.

- The time scale is a virtual clock whose output time t_k is the weighted mean value of the time $t_j(t_k)$ given by every clock C_j of the ensemble,

$$t_k = \sum_{j=1}^{N} w_j \times t_j(t_k) . \tag{7.8}$$

w_j is the weight of clock C_j,

$$\sum_{j=1}^{N} w_j = 1 . \tag{7.9}$$

- The time error of any clock C_j at a given epoch t_k is $\delta t_j(t_k)$,

$$\delta t_j(t_k) = t_j(t_k) - t_k . \tag{7.10}$$

- The time difference between the relay clock C_m and any clock C_j at the same epoch t_k is $\delta t_{jm}(t_k)$,

$$\delta t_{jm}(t_k) = \delta t_j(t_k) - \delta t_m(t_k) . \tag{7.11}$$

- The time error of any clock C_j $(1 \leq j \leq N)$ at the same epoch t_k is consequently

$$\delta t_j(t_k) = \delta t_\mathrm{m}(t_k) + \delta t_{jm}(t_k) . \tag{7.12}$$

- If all the time differences $\delta t_{jm}(t_k)$ are known at the epoch t_k, the time error of the relay clock can be computed, and consequently the value of the time scale

$$\sum_{j=1}^{N} w_j \times \delta t_{jm}(t_k) = \sum_{j=1}^{N} w_j \times [\delta t_j(t_k) - \delta t_m(t_k)] \tag{7.13}$$

$$= -\delta t_m(t_k) + \sum_{j=1}^{N} w_j \times \delta t_j(t_k) \tag{7.14}$$

$$= -\delta t_m(t_k) . \tag{7.15}$$

- The time error of every clock C_j can then be easily computed using (7.12).

The algorithm is then the following:

1. Epoch t_k:
   a) The time differences $\delta t_{jm}(t_k)$ are measured.
   b) The time error of the relay clock C_m is deduced from these measurements (7.15).
   c) The time error $\delta t_j(t_k)$ of every clock is deduced from these measurement by (7.12).
   d) The estimate $\widehat{\delta t_j}(t_{k+1})$ of the time error at epoch $t_{k+1} = t_k + \tau$ is calculated for every clock C_j ($j = m$ included) using (7.6) and the values of the parameters characterizing every clock.
2. Epoch t_{k+1}:
   a) The time differences $\delta t_{jm}(t_k)$ are measured.
   b) The time error of the relay clock C_m is deduced from these measurements (7.15).

   c) The time error $\delta t_j(t_k)$ of every clock is deduced from these measurement (7.12) and is compared to the time error estimated in step 1.
   d) The weight w_j of every clock C_j of the ensemble is adjusted by an algorithm that uses the difference between the time error $\delta t_j(t_{k+1})$ measured at epoch t_{k+1} and the time error $\widehat{\delta t_j}(t_{k+1})$ predicted.
   e) Steps b and c are repeated with the new values of the weights of the clocks.
   f) The estimate $\widehat{\delta t_j}(t_{k+2})$ of the time error at epoch $t_{k+2} = t_{k+1} + \tau$ is calculated for every clock C_j ($j = m$ included) using (7.6) and the values of the parameters characterizing every clock.
3. Step 2 is repeated at epochs $t_k + n\tau$.

The weight of a given clock is larger if the clock is predictable, i.e. its predicted error time is close to its measured one. The weight of a clock is often inversely proportional to the mean square of the prediction error.

Moreover, the weight of a clock can be canceled if its predicted error is larger than a given threshold, which may indicate a clock failure.

On the other hand, in order to avoid instability, the weight of any clock cannot be larger than a given limit. For instance, in the case of TAI, the maximal value of the weight of any clock is $w_{\max} = 2.5/N$, N being the number of clocks.

7.2 Comparison of Clocks

As shown in Sect. 7.1.2, the construction of a time scale relies on the measurement of the time difference between a particular clock of the ensemble and all the other ones. This measurement is equivalent to the synchronization of the clocks of the ensemble; if the time difference between every clock and the time scale is known, then all the clocks of the ensemble can be set to the correct time and are consequently synchronized.

A theoretical difficulty arises in the synchronization of clocks on the surface of the Earth, due to the fact that their reference system is not an inertial one but a rotating one [8].

7.2.1 Clock Synchronization in a Rotating Frame

All techniques used to synchronize clocks are variants of the so-called "Einstein clocks synchronization", which is a convention in special relativity describing how to synchronize clocks in different places in *an inertial frame.* This operation is necessary in order to measure and define the time coordinate of a given event.

The "Einstein clocks synchronization" process is the following: if clock C_{B}, located at point B is to be synchronized to clock C_{A} located at point A, a signal is sent at time t_1 from clock C_{A} to clock C_{B} and *immediately back*, (e.g. by the means of a mirror). Its arrival time back at clock C_{A} is t_2. The

Einstein synchronization convention then sets clock C_B so that the time of the signal reflection on clock C_B is $(t_1 + t_2)/2$. Information giving this value can be transmitted from A to B.

An equivalent process is the following: the time of arrival of the signal at clock C_B is $t_3 = t_1 + d/v$, where d is the distance between A and B and v the speed of the signal propagation. In the first process, the knowledge of t_1 and t_2 gives $l/v = (t_2 - t_1)/2$ and $t_3 = (t_1 + t_2)/2$, *assuming that the velocity of the signal is isotropic*. This process is transitive: If two clocks C_B and C_C (located at B and C, respectively) are synchronized to C_A, then C_B and C_C are synchronized. This results from the fact that in an inertial frame, the light velocity is constant and isotropic.

This property of transitivity applies in inertial frames only. Indeed, the process can be applied to rotating frames but is no longer transitive: If two clocks C_B and C_C are synchronized to C_A, then a time difference can exist between C_B and C_C. This result applies whatever the signal used to operate the synchronization may be. It applies even if the clock C_A is moved successively to the location of clock C_B and to the location of clock C_C. This phenomenon is the so-called Sagnac effect [109].

The following convention is consequently applied to define the synchronization of clocks on the rotating Earth [8].

A set of virtual synchronized clocks $C_V(i)$ keep the time coordinate in a virtual geocentric *inertial* frame. At every moment, each clock C_L of the rotating frame is synchronized to one of the virtual clocks, which is *located at exactly the same point* as the clock C_L. The time given by these synchronized clocks define the "coordinate time" and this process ensures that all the clocks on the rotating Earth give a common value.

In a practical way, if Einstein synchronization is to be applied to the clocks C_A and C_B located on the Earth, the duration of the signal travel from C_A to C_B must be corrected from the Sagnac effect. This is equivalent to saying that the speed of the signal is not isotropic[10].

7.2.2 One-way GPS Measurements

GPS maintains a time scale (the GPS time scale) linked to TAI and UTC time scales (see Sect. 7.1.1).

The relevant time information is transmitted by each GPS satellite (see Sect. 8.2.1) as

- a timing signal on a phase modulated L-band carrier,
- the difference between the time of that particular satellite and the GPS time, and
- the difference between the time scales GPS and UTC.

[10] This effect has produced a great deal of controversy, which is out of the scope of this book.

The most simple way to control the time indicated by a given clock is to compare it with the GPS time scale via the transmitted signal, in a one-way measurement [85]. This is the direct transposition of the Einstein clocks synchronization.

The source A (a GPS satellite) sends a time signal to the user, B through a transmission medium. The user's receiver compares the arrival time of the GPS signal to the time given by the clock to be characterized with a time interval counter. The arrival time of the GPS signal must be corrected for the propagation delay and the major limitation to the accuracy results from the estimation of this delay.

The delay d_{AB} over a transmission path is about 3.3 μs/km and the smallest distance between any GPS satellite and a point on the surface of the Earth is about 20 000 km, corresponding to a delay of 66 ms. Therefore, a high accuracy is required for the value of this delay; the physical locations of the two clocks (the clock in the satellite and the clock to be synchronized) must be known with an accuracy that is compatible with the accuracy of the synchronization. An error of 10 ns on time corresponds to an error on the distance between the two clocks of only 3 m.

Moreover, ionospheric and tropospheric delays must be taken into account.

1. The geometrical delay (the distance between the emitter and the receiver) is obtained from the receiver's location (which can be obtained from the GPS system) and from the broadcast satellite positions, which are only accurate to a few meters, and therefore limit the accuracy of the geometrical delay to about 10 ns. It is therefore not useful to determine the receiver position to better than one meter.
2. The ionosphere and the troposphere also have a significant impact on propagation delay. Ionospheric delays can range from a few nanoseconds at night in the vertical direction to hundreds of nanoseconds during the day at low elevation angles. The navigation message broadcast by the satellites contains a model of the ionospheric delay, but its accuracy may be rather poor.
   There are ways to improve the ionospheric correction with direct measurements using for, instance, the two-frequency method (see Sect. 8.2.2).
3. The excess delay introduced by the troposphere is in the range of tens of nanoseconds. It can generally be corrected to an uncertainty of a few nanoseconds.
4. Multipath error is produced if the signal, due to reflexions, reaches the receiver via more than one path. This may produce errors of a few tens of ns.
5. Hardware instabilities and inaccuracies generally produce errors of a few nanoseconds only.
6. The Sagnac delay is calculated from the positions and relative velocities of the emitter and of the receiver and can be of the order of 100 ns,

but uncertainty in this effect is small compared to other sources of error.

In conclusion, with a well designed receiver system, the user can usually obtain the time to better than 100 ns in a few minutes.

7.2.3 GPS Common-view

In this case, the two clocks to be compared are on the surface of the Earth and a given event E (a C/A code, for instance) is transmitted from a GPS satellite (see, for instance [85, 3]) and is received by the two clocks.

Since GPS satellites orbit the Earth within a 12 h period at an altitude (above the Earth's surface) of about 20 000 km on quasi-circular orbits inclined to the equator by about 55°, two receivers with longitudes differing by as much as 180° can receive the signal of *the same satellite* (Fig. 7.1).

Compared to the one-way measurement, this method has the following advantages:

1. Each station measures the apparent time of the same event, which is a given tick of the clock C_{E}. The stability and accuracy of the in-board clock that delivers the event E therefore has little effect; there is no need to know the date of event E, but only to compare the dates given by the receivers A and B when they receive this event. Any error in the time E is totally canceled.
2. Due to the geometry of the three points A, B and E, the perturbations that affect the propagation delay on the two paths EA and EB are correlated and their effects on the measurement of the time difference of C_{A} and C_{B} nearly cancel.

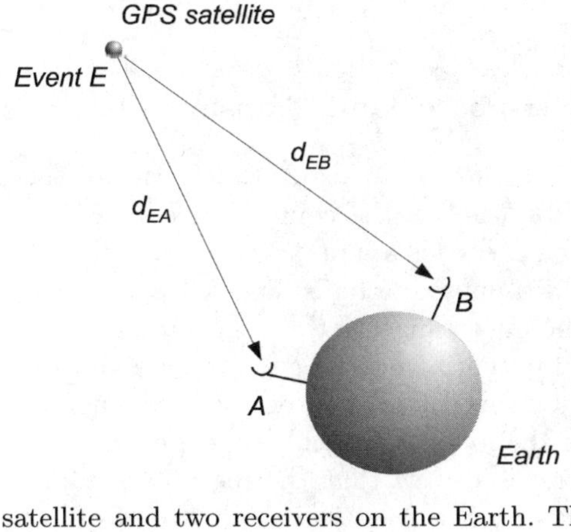

Fig. 7.1. A GPS satellite and two receivers on the Earth. The distance Earth–satellite is about four times the Earth's radius

In fact, the experimental results show that the accuracy of the method is at least ten times better than that of the one-way method.

The data resulting from the time difference measurements in the two stations A and B are interexchanged by the use of any convenient method (email, FTP, etc.).

The value given by stations A and B on measuring a common event produced at time t_E are, respectively, t_A and t_B given by

$$t_A = t_E + \delta_{AE}t + \delta t^p_{EA} , \quad (7.16)$$

$$t_B = t_E + \delta_{BE}t + \delta t^p_{EB} , \quad (7.17)$$

where t_E is the time of the clock C_E, δ_{AE} and δ_{BE} are the time errors of, respectively, C_A and C_B relative to the clock C_E, δt^p_{EA} and δt^p_{EB} are the propagation delays from the clock C_E to the clocks C_A and C_B, respectively.

The difference $t_A - t_B$ between the two results can easily be computed if the two stations can exchange data

$$t_A - t_B = (\delta_{AE} - \delta_{BE}) + (\delta t^p_{EA} - \delta t^p_{EB}) . \quad (7.18)$$

In this expression, $\delta_{AE} - \delta_{BE}$ is the time difference between the two clocks, i.e. the quantity to be measured.

The difference of the propagation delays, $\delta t^p_{EA} - \delta t^p_{EB}$ (including the delays of the receiving equipment), can be computed from the GPS data (the position of the receivers and of the satellite, etc.) and from a differential calibration of the receiving equipment. This differential calibration is made by carrying a portable GPS receiver between the two stations, which provides the value of the difference between the delays of the equipment. An absolute calibration of the equipment at each station is not required.

In conclusion, the accuracy of common-view time transfers is typically in the 1 to 10 ns range.

7.2.4 GPS Carrier-phase Time Transfer

Instead of the C/A code, the phase of the carrier of the signals emitted by the satellites may be used as the common event of the two receivers. Since the frequency of the carrier is about 1 000 times higher than the frequency of the C/A code, carrier-phase methods have a greater resolution. Nevertheless, the error due to the uncertainty of the delays remain the same. Therefore, in order to realize this potential advantage, it is necessary to analyze the received data after-the-fact, using the post-processed ephemeris of the satellites and detailed models of the ionosphere and troposphere.

Experimental results show that the method is capable of providing frequency comparisons with a fractional uncertainty of about 2×10^{-15} using one day of averaging [75, 76, 53].

7.2.5 Two-way Satellite Time and Frequency Transfer (TWSTFT)

Two-way satellite time and frequency transfer (TWSTFT) is a high precision method of time and frequency transfer, used principally for intercomparisons of local and international atomic time scales [37]. Intercontinental time transfer sessions can now achieve a precision of better than 1 ns [28, 102].

The principle of two-way time transfer is very simple [54, 62] (Fig. 7.2). In fact, it is a direct variant of the Einstein clock synchronization:

Clock C_A, located at station A sends at time t_1^A a time information (1 Hz pulses, for instance) to clock C_B located at station B. This time information is relayed by a satellite S. Station A receives at time t_2^A the time information sent by clock C_B and relayed by the same satellite S. In a symmetric way, clock C_B sends at time t_1^B time information to clock C_A and receives at time t_2^B the time information sent by clock C_A.

The time difference $t_2^\mathrm{A} - t_1^\mathrm{A}$ measured at station A by a time difference counter is Δt^A, while the corresponding time difference $t_2^\mathrm{B} - t_1^\mathrm{B}$ measured at station B is Δt^B

$$\Delta t^\mathrm{A} = t_1^\mathrm{B} - t_1^\mathrm{A} + \delta t_\mathrm{BA}^\mathrm{p} \,, \tag{7.19}$$

$$\Delta t^\mathrm{B} = t_1^\mathrm{A} - t_1^\mathrm{B} + \delta t_\mathrm{AB}^\mathrm{p} \,, \tag{7.20}$$

where $\delta t_\mathrm{BA}^\mathrm{p}$ and $\delta t_\mathrm{AB}^\mathrm{p}$ are, respectively, the transmission delays from B to A and from A to B.

The two stations exchange these results and compute the difference

$$\Delta t^\mathrm{A} - \Delta t^\mathrm{B} = 2(t_1^\mathrm{B} - t_1^\mathrm{A}) + (\delta t_\mathrm{BA}^\mathrm{p} - \delta t_\mathrm{AB}^\mathrm{p}) \,. \tag{7.21}$$

If the two transmission paths are reciprocal, $\delta t_\mathrm{BA}^\mathrm{p} - \delta t_\mathrm{AB}^\mathrm{p} = 0$ and the time difference between the two clocks is exactly

$$t_1^\mathrm{A} - t_1^\mathrm{B} = \frac{\Delta t_1^\mathrm{B} - \Delta t_1^\mathrm{A}}{2} \,. \tag{7.22}$$

In fact, the two paths are never completely reciprocal and an error subsists, which must be carefully evaluated [62, 102]. Moreover, due to the rotation of the Earth, the Sagnac effect (see Sect. 7.2.1) must be taken into account.

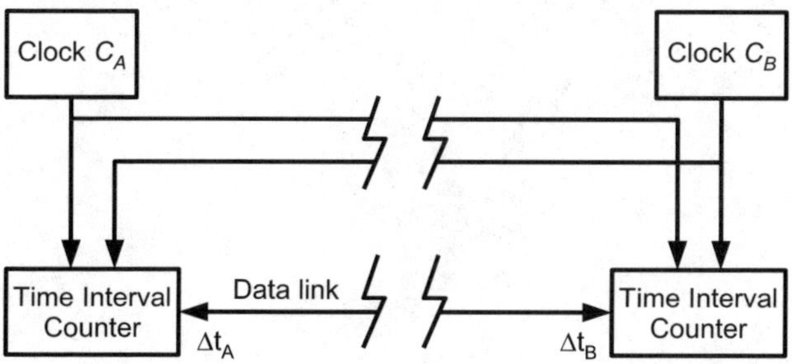

Fig. 7.2. The principle of the two-way technique

The paths of a TWSTFT go through a satellite S, which must be in common view of the two stations A and B. One path (from A to B) is the path $A \to S \to B$ and the other one (from B to A) is the path $B \to S \to A$ (Fig. 7.3).

The path from station A to station B, for instance, comprises

- a transmitter and an antenna at station A,
- an uplink from station A to the satellite S,
- a path through the satellite S,
- a downlink from the satellite S to station B, and
- an antenna and a receiver at station B.

The delays contributing to $\delta t^{\mathrm{p}}_{\mathrm{AB}}$ and $\delta t^{\mathrm{p}}_{\mathrm{BA}}$ are, therefore, the following:

1. The propagation delays from every station to the satellite and from the satellite to every station:
   a) d_{AS} – propagation from station A to the satellite S,
   b) d_{SA} – propagation from the satellite S to station A,
   c) d_{BS} – propagation from station B to the satellite S, and
   d) d_{SB} – propagation from the satellite S to station B.
   If the uplink frequency is different from the downlink frequency, $d_{\mathrm{AS}} \neq d_{\mathrm{SA}}$ and $d_{\mathrm{BS}} \neq d_{\mathrm{SB}}$, due to the dispersion in the ionosphere, and the difference $(d_{\mathrm{BS}} - d_{\mathrm{AS}}) + (d_{\mathrm{SA}} - d_{\mathrm{SB}})$ does not cancel and produces the ionospheric delay.
   This delay can be computed from an evaluation of the index of refraction of the signal in the ionosphere [102]. It depends on the frequencies of the uplink and of the downlink, of the elevation of the satellite and of the

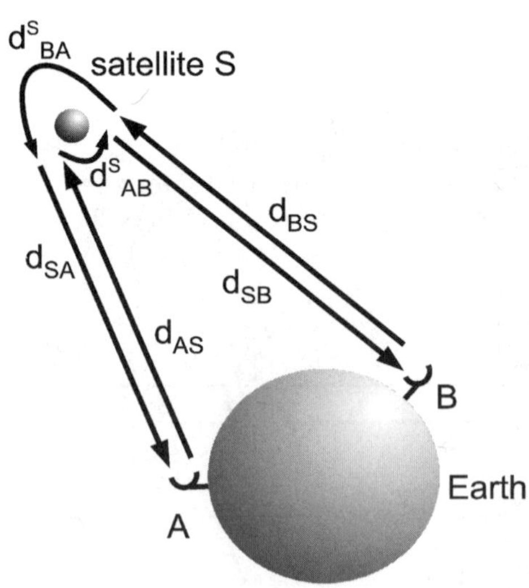

Fig. 7.3. Paths of TWSTFT

various factors that affect the electron density in the ionosphere (day and night, solar activity, etc. [68]). The non-reciprocity due to this effect is smaller than 100 ps [102].

2. The delays produced by the satellite electronic transmitter:
   a) d^{S}_{AB} – the delay produced on the signal sent by station A and transmitted to station B, and
   b) d^{S}_{BA} – the delay produced on the signal sent by station B and transmitted to station A.

   If the two path directions go through the same transponder at the same frequency, which is possible using the code division multiple access (CDMA) mode with PN signals[11], the two delays are equal: $d^{\mathrm{S}}_{\mathrm{AB}} = d^{\mathrm{S}}_{\mathrm{BA}}$.
3. The delays produced by the electronic equipment on the Earth stations:
   a) $d^{\mathrm{T}}_{\mathrm{A}}$ – the delay produced by the transmitter of station A,
   b) $d^{\mathrm{R}}_{\mathrm{A}}$ – the delay produced by the receiver of station A,
   c) $d^{\mathrm{T}}_{\mathrm{B}}$ – the delay produced by the transmitter of station B, and
   d) $d^{\mathrm{R}}_{\mathrm{B}}$ – the delay produced by the receiver of station B.
4. The Sagnac effect. This effect, which appears in a rotating frame, is described in Sect. 7.2.1. Its consequence is that the time required for an electromagnetic signal to travel from a station A to a station B is different from the time required to travel from B to A, depending on whether the signal is moving against or with the rotation of the reference system. The Sagnac effect introduces the asymmetry $d^{\mathrm{Sa}}_{\mathrm{AB}}$ between the two paths $A \to M$ and $B \to M$. Nevertheless, this asymmetry can easily be computed to a good accuracy (better than 1 ns) with only a poor knowledge of the positions of A, B and S [54].

The term $\delta t^{\mathrm{p}}_{\mathrm{BA}} - \delta t^{\mathrm{p}}_{\mathrm{AB}}$ is then

$$\begin{aligned}\delta t^{\mathrm{p}}_{\mathrm{BA}} - \delta t^{\mathrm{p}}_{\mathrm{AB}} &= (d_{\mathrm{BS}} - d_{\mathrm{AS}}) + (d_{\mathrm{SA}} - d_{\mathrm{SB}}) + (d^{\mathrm{S}}_{\mathrm{BA}} - d^{\mathrm{S}}_{\mathrm{AB}}) \\ &\quad + (d^{\mathrm{T}}_{\mathrm{B}} - d^{\mathrm{T}}_{\mathrm{A}}) + (d^{\mathrm{R}}_{\mathrm{A}} - d^{\mathrm{R}}_{\mathrm{B}}) \,. \end{aligned} \tag{7.23}$$

The delays introduced by the Earth equipment can be calibrated using a mobile Earth station M as follows:

1. If the mobile Earth station M is collocated with the station A and uses the same clock C_{A} as station A:
   - the Sagnac effect cancels,
   - the time difference between the two clocks under test cancels, since it is the same clock, and

[11] In such signals, information is represented by pseudo-random noise sequences. If the sequences representing the two signals have low cross-correlation, they can be handled at the same time and in the same range of frequencies by the satellite transponder without interference. CDMA is the common platform on which 3G technologies are built.

– the difference in the measurements in A and M is consequently

$$\Delta t_1^{\mathrm{A}} - \Delta t_1^{\mathrm{M}} = 2(t_1^{\mathrm{M}} - t_1^{\mathrm{A}}) + (\delta t_{\mathrm{MA}}^{\mathrm{P}} - \delta t_{\mathrm{AM}}^{\mathrm{P}}) \tag{7.24}$$

$$= \delta t_{\mathrm{MA}}^{\mathrm{P}} - \delta t_{\mathrm{AM}}^{\mathrm{P}} \tag{7.25}$$

with

$$\delta t_{\mathrm{MA}}^{\mathrm{P}} = d_{\mathrm{MS}} + d_{\mathrm{SA}} + d_{\mathrm{MA}}^{\mathrm{S}} + d_{\mathrm{M}}^{\mathrm{T}} + d_{\mathrm{A}}^{\mathrm{R}} \,, \tag{7.26}$$

$$\delta t_{\mathrm{AM}}^{\mathrm{P}} = d_{\mathrm{AS}} + d_{\mathrm{SM}} + d_{\mathrm{AM}}^{\mathrm{S}} + d_{\mathrm{A}}^{\mathrm{T}} + d_{\mathrm{M}}^{\mathrm{R}} \,. \tag{7.27}$$

Therefore,

$$\begin{aligned} \Delta t_{\mathrm{A}} - \Delta t_{\mathrm{M}} &= \delta t_{\mathrm{MA}}^{\mathrm{P}} - \delta t_{\mathrm{AM}}^{\mathrm{P}} \\ &= d_{\mathrm{MS}} - d_{\mathrm{AS}} + d_{\mathrm{SA}} - d_{\mathrm{SM}} \\ &\quad + d_{\mathrm{MA}}^{S} - d_{\mathrm{AM}}^{S} + d_{\mathrm{M}}^{\mathrm{T}} - d_{\mathrm{A}}^{\mathrm{T}} + d_{\mathrm{A}}^{\mathrm{R}} - d_{\mathrm{M}}^{\mathrm{R}} \\ &= d_{\mathrm{MA}}^{\mathrm{S}} - d_{\mathrm{AM}}^{\mathrm{S}} + d_{\mathrm{M}}^{\mathrm{T}} - d_{\mathrm{A}}^{\mathrm{T}} + d_{\mathrm{A}}^{\mathrm{R}} - d_{\mathrm{M}}^{\mathrm{R}} \,. \end{aligned} \tag{7.28}$$

2. If the mobile Earth station M is now collocated with the station B and uses the same clock C_{B} as station B, the difference in the measurements in B and M is

$$\begin{aligned} \Delta t_3^{\mathrm{B}} - \Delta t_3^{\mathrm{M}} &= \delta t_{\mathrm{MB}}^{\mathrm{P}} - \delta t_{\mathrm{BM}}^{\mathrm{P}} \\ &= d_{\mathrm{MS}} - d_{\mathrm{BS}} + d_{\mathrm{SB}} - d_{\mathrm{SM}} \\ &\quad + d_{\mathrm{MB}}^{\mathrm{S}} - d_{\mathrm{BM}}^{\mathrm{S}} + d_{\mathrm{M}}^{\mathrm{T}} - d_{\mathrm{B}}^{\mathrm{T}} + d_{\mathrm{B}}^{\mathrm{R}} - d_{\mathrm{M}}^{\mathrm{R}} \\ &= d_{\mathrm{MB}}^{\mathrm{S}} - d_{\mathrm{BM}}^{\mathrm{S}} + d_{\mathrm{M}}^{\mathrm{T}} - d_{\mathrm{B}}^{\mathrm{T}} + d_{\mathrm{B}}^{\mathrm{R}} - d_{\mathrm{M}}^{\mathrm{R}} \,. \end{aligned} \tag{7.29}$$

3. Comparing the two sessions gives

$$\begin{aligned} (\Delta t_1^{\mathrm{A}} - \Delta t_1^{\mathrm{M}}) - (\Delta t_3^{\mathrm{B}} - \Delta t_3^{\mathrm{M}}) &= d_{\mathrm{MA}}^{\mathrm{S}} - d_{\mathrm{MB}}^{\mathrm{S}} + d_{\mathrm{BM}}^{\mathrm{S}} - d_{\mathrm{AM}}^{\mathrm{S}} \\ &\quad + d_{\mathrm{B}}^{\mathrm{T}} - d_{\mathrm{A}}^{\mathrm{T}} + d_{\mathrm{A}}^{\mathrm{R}} - d_{\mathrm{B}}^{\mathrm{R}} \\ &= d_{\mathrm{B}}^{\mathrm{T}} - d_{\mathrm{A}}^{\mathrm{T}} + d_{\mathrm{A}}^{\mathrm{R}} - d_{\mathrm{B}}^{\mathrm{R}} \,. \end{aligned} \tag{7.30}$$

In (7.30), it has been supposed that the satellite transponder is reciprocal, by using CDMA mode, for instance. This equation gives the value of the global effect of the equipment of the Earth stations.

A detailed analysis of the time transfer instabilities produced by these parameters instabilities is carried out in [102]. With the exception of the Sagnac effect, and assuming reciprocity of the physical paths, the precise knowledge of the positions of the satellite and of the Earth stations is not necessary.

8 Navigation and Localization: Global Positioning System

Accurate and stable timing signals emitted by well localized stations allow any receiver to determine its precise localization by a triangulation computation.

The idea of the global positioning system is to obtain worldwide coverage by putting the emitters in satellites orbiting around the Earth[1]. The drawback is the fact that the emitters are not fixed in a given position, which is a source of uncertainty and implicates the need to maintain a knowledge as accurate as possible of their ephemeris.

The global positioning system, GLONASS and Galileo provide or will provide high accuracy localization and time transfer to any user. Only GPS is now fully operational and will be discussed here. GPS is based on a constellation of at least[2] 24 satellites that broadcast precise radio timing signals. Its official name is NAVSTAR GPS (navigation signal timing and ranging GPS).

Any observer that can receive and process the information broadcast by at least four of these satellites can accurately determine its location (longitude, latitude and altitude) in real time as well as the value of UTC.

This system was developed and is maintained by the United States Department of Defense. The satellite constellation is managed by the 50th Space Wing at Schriever Air Force Base. The cost of maintaining the system, including the replacement of aging satellites, is approximately US$ 400 million per year. GPS is now available for free use in civilian applications anywhere on the Earth.

The wide area augmentation system (WAAS), available since August 2000, increases the accuracy of GPS signals to within 2 meters. Moreover, the accuracy can be improved further, to about 1 cm over short distances, using techniques such as differential GPS (DGPS, see Sect. 8.2.3).

GPS has numerous applications in navigation on land, sea, and air around the world, and is an important tool for map-making and land surveying. It also provides an extremely precise time reference, giving access to UTC, required for telecommunications and some scientific research, including the study of troposphere, ionosphere and earthquakes (see Sect. 9).

---

[1] Satellites are often called "space vehicles" – SV – in the specialized literature.

[2] The nominal GPS operational constellation consists of 24 satellites but they are often more numerous as new ones are launched to replace older satellites.

At the end of 2005, the first in a series of next-generation GPS satellites was added to the constellation, offering several new capabilities, including a second civilian GPS signal called L2C for enhanced accuracy and reliability.

8.1 The GPS Infrastructure

The system relies on three main parts (called segments).

8.1.1 The Space Segment

The space segment consists of the GPS satellites and their equipment. The nominal GPS Operational Constellation is made of 24 satellites that orbit the Earth in 12 hours. The orbit altitude (20 000 to 25 000 km) is such that the satellites repeat the same track and configuration over any point of the Earth approximately each 24 hours.

The satellites orbit the Earth in six orbital planes (with nominally four satellites in each), equally spaced (each plane is deduced from another one by a rotation of 60° along the polar axis), and inclined at about 55° degrees with respect to the equatorial plane. The arrangement of the orbits is shown in Fig. 8.1.

This disposition allows any user in the vicinity of the Earth's surface to receive between 5 and 12 satellites at any time.

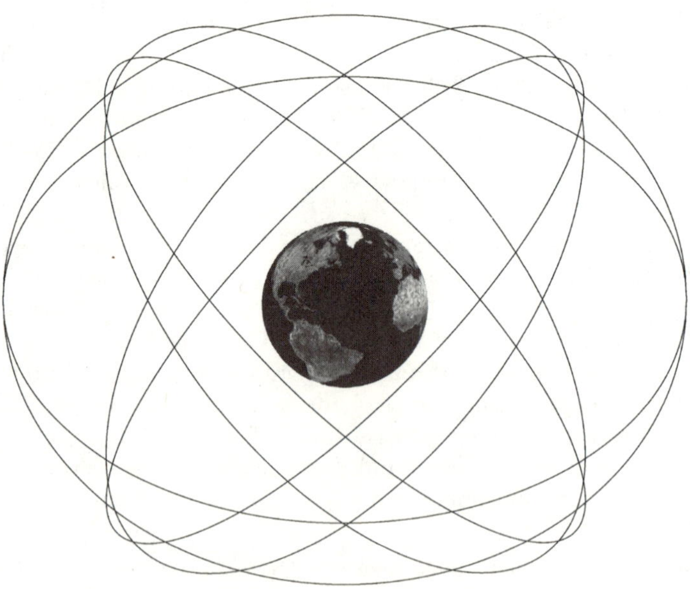

Fig. 8.1. Orbits of the GPS satellites

The operational satellites belong to different generations (Block II, Block IIA, Block IIR and Block IIR-M) and do not have the same equipment, but all of them contain atomic clocks, emitters and receivers.

The Block II satellites were launched from February 1989 through October 1990. They were designed to provide 14 days of operation without contact from the control segment (see below). One satellite from Block II is still operational.

The Block IIA satellites were launched from November 1990 through November 1997. They were designed to provide 180 days of operation without contact from the control segment, but during this long period of autonomy, the accuracy of the navigation message was degraded. Fifteen satellites from block IIA are operational.

The design life of the Block II/IIA satellites is 7.3 years; each contain two cesium atomic clocks and two rubidium atomic clocks and have selective availability (SA) and anti-spoof (A-S) capabilities (see below).

The Block IIR satellites are designed to provide at least 14 days of operation without contact from the control segment and up to 180 days of operation when operating in the autonomous navigation (AUTONAV) mode. Full accuracy will be maintained using a technique of ranging and communication between the Block IIR satellites themselves. Cross-link ranging will be used to estimate and update the parameters in the navigation message of each Block IIR satellite without contact from the control segment.

The design life of the Block IIR satellite is 7.8 years; each contains three Rb atomic clocks and have SA and A-S capabilities (see below). Launching of the Block IIR satellites began in January 1997, thirteen satellites from block IIR are operational.

The first GPS IIR-M satellite was launched on 26 September 2005. This new series of satellites will broadcast two new military signals and a second civil signal (L2C). Detailed information on the GPS satellites is given in [121, 122] and [111].

8.1.2 The Control Segment

Five stations around the world monitor the data given by the satellites. These stations are located on the following islands: Hawaii (Pacific Ocean), Kwajalein (Pacific Ocean), Ascension Island (Atlantic Ocean), Diego Garcia (Indian Ocean). The master station is located at Colorado Springs (Schriever Air Force Base, 50th Space Wing, 1st Space Operations Squadron).

The monitor stations are equipped with atomic clocks and GPS receivers located on geodesic points whose coordinates are known with great accuracy. The data obtained by these receivers allow the calculation of the position of the satellites and the time difference between the clock inboard the satellites and GPS time.

The master control station processes the observations of the monitor stations and sends updates to the satellites through the stations at Ascension

Island, Diego Garcia and Kwajalein. As a result, the atomic clocks inboard the satellite are synchronized to GPS time to within one microsecond and the ephemeris of the internal orbital model of the satellite is adjusted in order to match the observations of the satellites from the ground. These corrected data are incorporated in the message emitted by the satellite.

8.1.3 The User Segment

The GPS user segment consists of GPS receivers, which are passive and convert the data received from the satellites into the four coordinates: latitude, longitude, altitude and time. Four satellites are required to compute these four dimensions of X, Y, Z (position) and time.

The following applications are possible:

- Navigation is the primary function of GPS. Navigation receivers are made for aircraft, ships, ground vehicles, and for hand carrying as well.
- More precise positioning is possible using a differential scheme where the position of the operating receiver is compared to the precisely known position of a reference receiver. Some of the errors limiting the accuracy are eliminated in this differential scheme, which is applied in surveying, geodetic control, and plate tectonic studies, for example.
- Time and frequency dissemination, based on the precise clocks on board the satellites and controlled by the monitor stations (GPS time scale, see Sect. 7.1.1) is another use for GPS. Astronomical observatories, telecommunications facilities and laboratory standards can be set to precise time signals or controlled to accurate frequencies by special purpose GPS receivers. Individuals can also access a precise time with a hand-held receiver.

8.2 GPS Data Processing

8.2.1 Data Transmitted by the Satellites

The GPS satellites continuously emit a numerous amount of accurate and necessary information to the receiver in order to calculate its position and the time. The coding, formats and protocols are described below.

Media and Coding

The data are transmitted by the satellites using the spread spectrum technique[3]: the binary useful message is scrambled by a pseudo-random noise (PRN) sequence before modulating the carrier.

[3] In the spread spectrum technique, a signal is transmitted on a bandwidth much larger than the frequency content of the original information. It employs direct

Two microwave carriers named L_1 and L_2 (Fig 8.2) and two PRN sequences are involved in the satellite emission:

1. Carrier L_1 carries the navigation message whose spectrum has been spread by:
   - the C/A-code (coarse acquisition code), which allows the so-called standard positioning service (SPS),
   - the P-code (precise code), which allows the so-called precise positioning device (PPS).

   Its frequency – 1575.42 MHz – results from the multiplication of the base frequency $f_0 = 10.23$ MHz by a multiplying factor of 154.
2. Carrier L2 carries the navigation message whose spectrum has been spread by:
   - the same P-code as L_1;

   its frequency – 1227.60 MHz – results from the multiplication of the base frequency $f_0 = 10.23$ MHz by a multiplying factor of 120.

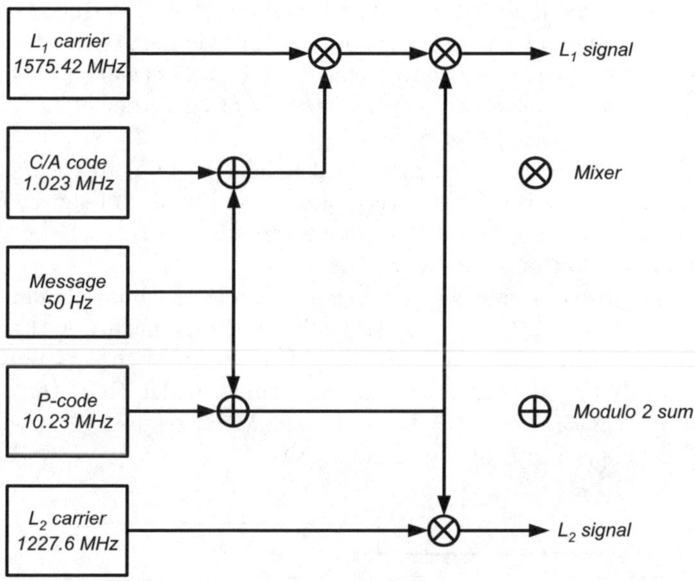

Fig. 8.2. L_1 and L2 signals emitted by the GPS satellites

---

sequence, frequency hopping or a hybrid of these. It is extensively used for wireless Ethernet links, digital cellular telephony, etc. It decreases the interference to other receivers and achieves privacy. The spread spectrum technique generally makes use of a sequential noise-like signal structure (pseudo random noise) to spread the band information signal over a wider band of frequencies. The receiver correlates the received signals to retrieve the original information. The receiver must be aware of the sequence used in order to retrieve the information.

The same P-code is transmitted on two different frequencies, which allows precise positioning service equipped receivers to measure the ionospheric delay (from the delay difference between the two frequencies) and to obtain a better precision than that obtained from the model emitted by the satellite.

Until 2000, the so-called selective availability intentionally degraded the SPS accuracy. In 2000, selective availability was turned off by US presidential order, giving all GPS receivers the potential accuracy of 15 meters without the use of signal correction.

In the anti-spoofing (AS) mode of operation, the P-code is encrypted into the Y-code. The encrypted Y-code requires a classified AS module for each receiver channel and can be used only by authorized users with cryptographic keys.

Construction of the signals L_1 and L_2

1. The satellite navigation message is a continuous 50 bits/s data stream that describes the GPS satellite orbits, clock corrections and other system parameters. The detailed format of this message is described below. The envelope of the Fourier transform of this signal is the Fourier transform of an elementary bit, i.e. is proportional to $1/T_{\mathrm{m}} \times \mathrm{sinc}(\pi\nu T_{\mathrm{m}})$. The width of the spectrum of this signal is consequently of the order of $1/T_{\mathrm{m}} = 50\,\mathrm{Hz}$ and its amplitude is proportional to T_{m}.
2. This data stream is added modulo 2 (exclusive OR) to the PRN C/A-code and to the PRN P-code (Fig. 8.3). The bit rate of the C/A-code is exactly 20 460 times the bit rate message, the bit rate of the P-code is exactly 204 600 times the bit rate message.
   The result of these additions are two sequences whose elementary bits are, respectively, 20 460 times and 204 600 times narrower than that of the message sequence. Consequently, the spectra of these sequences are, respectively, 20 460 times and 204 600 times wider than the spectrum of the message (i.e. 1.023 MHz and 10.23 MHz, respectively) while their amplitude is reduced by the same factor.

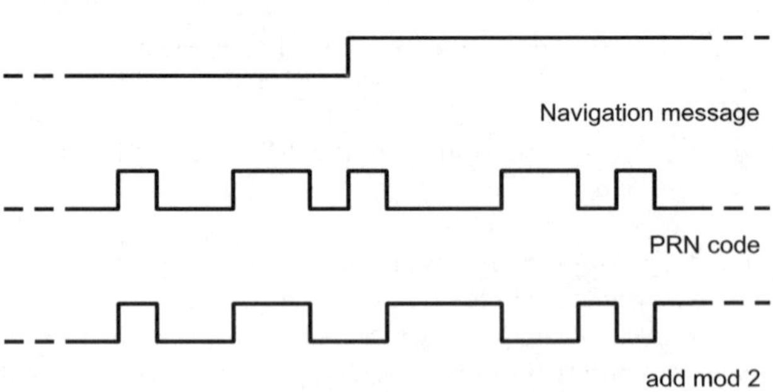

Fig. 8.3. Modulo 2 (XOR) addition of the navigation message and the C/A-code and the P-code

It is impossible to read the original message in any of the two resulting sequences, unless the codes (C/A-code and P-code) that have been used to spread its spectrum are known. On the contrary, if the C/A-code or the P-code is known and synchronized, it is easy to retrieve the navigation message by simple logic operation. In fact, an XOR gate can deliver the navigation message; see Fig. 8.4. The user receiver must independently generate and synchronize with the satellite transmitted C/A-code and perform modulo 2 addition in order to decode and interpret the navigation message.

3. These resulting sequences modulate the phase of each carrier; any transition $0 \rightarrow 1$ or $1 \rightarrow 0$ shifts the phase by π.
   One elementary bit $0 \rightarrow 1 \rightarrow 0$ occurring at time t_0 produces, for instance, the phase variation of the carrier

$$\phi(t) = 2\pi\nu_0 t + \pi\Pi(t - t_0) , \tag{8.1}$$

   where $\Pi(t)$ is the square function of width T_c, the period of the PRN-code.
   The Fourier transform $X(\nu)$ of the signal $x(t) = x_0 \exp[j\phi(t)]$ emitted by the satellite is then

$$\begin{aligned} X(\nu) &= \int_{-\infty}^{\infty} x(t)\exp(-2j\pi\nu t)\mathrm{d}t \\ &= \int_{-\infty}^{\infty} x_0 \exp[-2j\pi(\nu-\nu_0)t]\exp(j\pi\Pi(t-t_0))\mathrm{d}t \\ &= \int_{-\infty}^{t_0-T_c/2} x_0 \exp[-2j\pi(\nu-\nu_0)t]\mathrm{d}t \\ &\quad + \int_{t_0+T_c/2}^{\infty} x_0 \exp[-2j\pi(\nu-\nu_0)t]\mathrm{d}t \\ &\quad + \int_{t_0-T_c/2}^{t_0+T_c/2} x_0 \exp[-2j\pi(\nu-\nu_0)t]\exp(j\pi)\mathrm{d}t . \end{aligned}$$

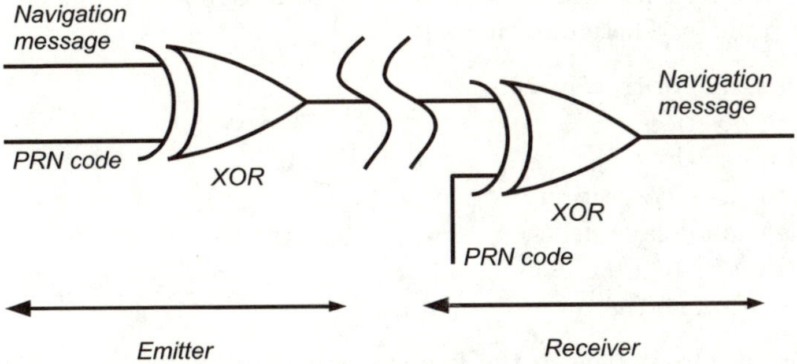

Fig. 8.4. Easy retrieval of the navigation message

Since

$$x_0 \exp[-2j\pi(\nu - \nu_0)t] \exp(j\pi) = -x_0 \exp[-2j\pi(\nu - \nu_0)t] ,$$

the Fourier transform is

$$\begin{aligned} X(\nu) = & \int_{-\infty}^{\infty} x_0 \exp[-2j\pi(\nu - \nu_0)t] \mathrm{d}t \\ & - 2 \int_{t_0 - T_c/2}^{t_0 + T_c/2} x_0 \exp[-2j\pi(\nu - \nu_0)t] \mathrm{d}t \\ = & \, x_0 \delta(\nu - \nu_0) \\ & - 2x_0 T_c \exp[-2j\pi(\nu - \nu_0)t_0] \mathrm{sinc}[\pi(\nu - \nu_0)T_c] . \end{aligned} \tag{8.2}$$

This result shows that

- the width is of the order of $1/T_c$ Hz and
- the amplitude of the spectrum of the signal emitted by the satellite is proportional to T_c (T_c is the period of the PRN-code, C/A-code or P-code).

The L_1 signal is consequently spread over a frequency range of about 1 MHz, while the L_2 wave is spread over about 10 MHz.

Format and Protocol

The signal specifications (for the standard positioning service) are detailed in [119] and [120].

The Structure of the Navigation Message

The navigation message is a continuous 50 bits/second data stream. Each message is 1 500 bits long and takes 30 seconds to be transmitted. The start of any message frame is precisely on the minute and half minute, according to the clock of the satellite.

Each GPS satellite provides data required to support the position determination process. These data include:

- satellite time of transmission,
- satellite position,
- satellite health,
- satellite clock correction,
- propagation delay effects,
- time transfer to UTC,
- constellation status.

This information is given in the five subframes, each 300 bits long, of every frame.

1. Subframe 1: GPS week number, satellite accuracy and health, satellite clock correction terms.
2. Subframes 2 and 3: Ephemeris parameters.
3. Subframe 4: Almanac and health data for satellites 25–32, special messages, satellite configuration flags, ionospheric and UTC data.
4. Subframe 5: Almanac and health data for satellites 1–24, almanac reference time, week number.

The complete information "almanac and health data for satellites 25–32, special messages, satellite configuration flags, ionospheric and UTC data" (subframe 4) is spread over 25 successive values (pages) of this subframe. In a similar way, the complete information "almanac and health data for satellites 1–24, almanac reference time, week number" (subframe 5) is given by 25 successive values (pages) of this subframe.

Each subframe contains 10 words of 30 bits each. The first word of each subframe is called a telemetry (TLM) word and the second one is called a handover word (HOW). It is followed by eight data words giving the information mentioned above. Each word in each frame contains parity. The first 8 bits in the telemetry word contain a sync pattern, used by the receiver to synchronize itself with the navigation message.

Time Information

Satellite Time The satellite time is connected to the GPS time scale (see Sect. 7.1.1), which is established by the control segment and is used as the primary time reference for all GPS operations. GPS time is referenced to a UTC (as maintained by the US Naval Observatory), the origin being defined as midnight on the night of 5 January 1980/morning of 6 January 1980.

GPS time differs from UTC because GPS time is a continuous time scale, while UTC is corrected when necessary with a leap second (see Sect. 7.1.1). There is also an inherent drift rate between the UTC and GPS time scales, but the GPS time scale is maintained to be within one microsecond of UTC (modulo one second). The navigation data contains the requisite data for relating GPS time to UTC. The largest unit used in stating GPS time is one week (604 800 seconds).

In each satellite, a 1.5 second epoch derived from the inboard clocks provides the unit for precisely counting and communicating time. Time stated in this manner is referred to as a Z-count. This Z-count is provided to the user as a 29-bit binary number consisting of two parts as follows:

1. The binary number represented by the 19 least significant bits of the Z-count is referred to as the time of week (TOW) count and is defined as being equal to the number of units (of 1.5 seconds) that have occurred since the transition from the previous week. The range of the TOW-count is from 0 to 403 199 units (i.e. 604 800 s, which corresponds to one week)

and is reset to zero at the end of each week. The TOW-count's zero state is defined as the start of the present week. This epoch occurs at (approximately) UTC midnight Saturday night/Sunday morning, the difference depending on the number of leap seconds of UTC (see Sect. 7.1.1).
In order to aid rapid ground lock-on, a truncated version of the value of the TOW-count *at the start of the next subframe*, consisting of its 17 most significant bits, is contained in the hand-over word (HOW, the second word of each subframe of the navigation message); the actual TOW-count is converted from the HOW-message by a multiplication of $2^2 = 4$. The time unit being 1.5 s, the time (in seconds) deduced from this truncated Z-count must be multiplied by 6.

2. The sequential number of the present GPS week (modulo 1024) is given by the ten most significant bits of the Z-count. The range of this count is consequently from 0 to 1 023. The GPS time scale was initialized on (UTC) 0 h 0 m 0 s 6 January 1980 (MJD 44 244) and was then set equal to UTC. This epoch is the origin of the week count. At the expiration of GPS week number 1 023 (MJD 51 412, 22 August 1999), the GPS week number has rollover to zero (0). Users must account for the previous 1 024 weeks in conversions from GPS time to a calendar date.
The third word of the first subframe contains the ten bits giving the sequential number of the present GPS week (week n°).

GPS Time Subframe 1 (Bits 9 through 24 of word 8, bits 1 through 24 of word 9, and bits 1 through 22 of word 10) contains the parameters allowing the satellite clock to be corrected relative to the GPS time scale.

UTC Page 18 of subframe 4 includes the parameters needed to relate GPS time to UTC, and a notice to the user regarding the scheduled future or recent past value of the delta time due to leap seconds (DtLSF), together with the week number (WNLSF) and the day number (DN) at the end of which the leap second becomes effective.

Ephemeris Information

The ephemeris data transmitted are the following:

M_0	Mean anomaly at reference time
Δn	Mean motion difference from computed value
e	Eccentricity
$(A)^{1/2}$	Square root of the semimajor axis
$(OMEGA)_0$	Longitude of ascending node of orbit plane at weekly epoch
i_0	Inclination angle at reference time
ω	Argument of perigee
OMEGADOT	Rate of right ascension
IDOT	Rate of inclination angle
C_{uc}	Amplitude of the cosine harmonic correction term to the argument of latitude

C_{us}	Amplitude of the sine harmonic correction term to the argument of latitude
C_{rc}	Amplitude of the cosine harmonic correction term to the orbit radius
C_{rs}	Amplitude of the sine harmonic correction term to the orbit radius
C_{ic}	Amplitude of the cosine harmonic correction term to the angle of inclination
C_{is}	Amplitude of the sine harmonic correction term to the angle of inclination
toe	Reference time ephemeris
IODE	Issue of data (ephemeris)

The almanac is a subset of the clock and ephemeris data, with reduced precision.

Updating the Navigation Message

The control segment typically updates the navigation message of each satellite every 24 hours by uploading new values. In an upload the master control station (MCS) sends to the satellite all of the data that the satellite will transmit during the next 24 hours. When a satellite begins transmitting a new data set this is called a cutover. The first cutover after an upload may occur at any time of the hour, but subsequent cutovers only occur precisely on hour boundaries. At any time, a receiver can ensure that it is using up-to-date ephemeris and clock data for the satellites in view due to the issue of data clock (IODC), provided in subframe 1 and the issue of data, ephemeris (IODE), provided in both subframes 2 and 3 values. Whenever these three terms do not match with those memorized by the receiver, a data set cut-over has occurred and new data must be collected. Moreover, any typical receiver is all-in-view, which means that it tries to track all satellites in view continuously.

8.2.2 Data Processing by the Receiver

In a first step, the receiver must collect the data transmitted by the satellites. Then, it must process these data in order to obtain the required information, such as its position.

Reading the Data

Any GPS receiver must retrieve some raw data from the received message in order to compute its position.

The PRN Codes: Pseudo-Range

The value of the raw pseudo-range (pseudo-distance between the receiver and a given satellite) is obtained by comparing the (C/A or P) PRN sequence received by the receiver with the same signal generated (and synchronized) by the receiver, which must consequently be aware of the code emitted by each satellite (see Fig. 8.5):

1. The C/A-code is a repeating 1.023 MHz Code. Since its length is 1023 bits, it repeats every 1 ms. The time difference between the local code and the received code is consequently measured with an ambiguity of an integer number of ms, corresponding to an integer multiple of 299 792.458 m in the pseudo-range.
   In fact, since the minimal vertical distance between any satellite and the Earth's surface is 20 000 km, the minimal time delay for a receiver located on the Earth surface is 66.7 ms. This ambiguity can easily be raised due to an estimated value (with an error smaller than 300 km) of the coordinates of the receiver.
2. The P-code is also a repeating code, but its length is of the order of one week. There is consequently no ambiguity in the time delay measurement.

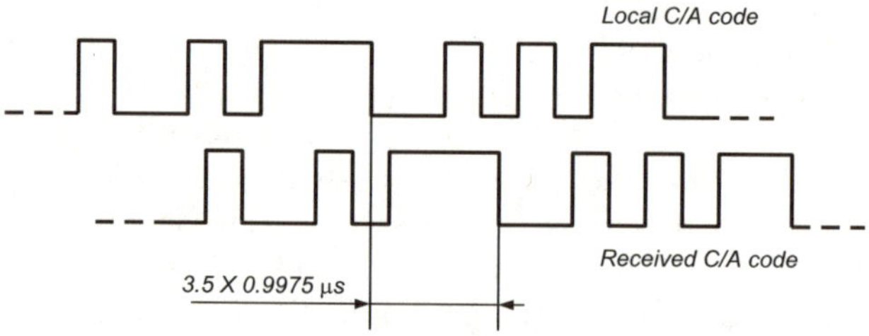

Fig. 8.5. Measurement of the raw pseudo-range (modulo 299 792.458 m)

The Navigation Message

The data transmitted by the navigation message are described in Sect. 8.2.1.

Knowing the PRN-code spreading its spectrum, it is easy for the receiver to detect this message (see Sect. 8.2.1).

It is to be noted that an elementary message is 30 s long and that 25 successive messages (each one is called a page) are necessary to transmit the full ephemeris and almanac data.

Pseudo-Range Calculation

The receiver can deduce its position from the messages it receives from at least four satellites by a mere triangulation computation. The observed pseudo-range PR (pseudo-distance between the receiver and the satellite) is deduced from the propagation time of the signal emitted by the satellite:

$$PR_{\mathrm{m}} = c(t_{\mathrm{r}} - t_{\mathrm{t}}) , \tag{8.3}$$

where c is the speed of light, t_{r} is the time in which the ranging measurement was received, measured by the clock of the receiver, t_{t} is time in which the ranging signal was transmitted from the satellite. These times must first be corrected from the satellite clock bias and propagation delays.

1. Satellite clock correction. The polynomial whose coefficients a_{f0}, a_{f1} and a_{f2} and reference time t_{oc} are transmitted in subframe 1 (bits 9 through 24 of word 8, bits 1 through 24 of word 9, and bits 1 through 22 of word 10) allows the user to determine the effective satellite PRN code phase offset referenced to the phase center of the satellite antennas ($\Delta t_{\mathrm{sv}}^{\mathrm{clock}}$) with respect to GPS system time (t) at the time of data transmission

$$\Delta t_{\mathrm{sv}}^{\mathrm{clock}} = a_{\mathrm{f0}} + a_{\mathrm{f1}}(t - t_{\mathrm{oc}}) + a_{\mathrm{f2}}(t - t_{\mathrm{oc}})^2 . \tag{8.4}$$

   This correction accounts for the deterministic satellite clock error (bias, drift and aging), as well as for the satellite group delay and mean differential group delay (the group delay is defined as the delay between the L-band radiated output of a specific satellite – measured at the antenna phase center – and the output of its on-board frequency source).
2. Satellite differential group delay T_{GD} [31], the difference between the group delays of the waves L_1 and L_2. The value of this delay is given by the bits 17 through 24 of word seven of subframe 1 data.
3. Relativistic correction Δt_{r}. This correction is deduced from the ephemeris parameters given in subframes 2 and 3

$$\Delta t_{\mathrm{r}} = FeA^{1/2} \sin(E_{\mathrm{k}}) , \tag{8.5}$$

   where e is the eccentricity, $A^{1/2}$ is the square root of the semimajor axis and E_{k} is deduced from Kepler's equation for eccentric anomaly.

$$F = -\frac{2\mu^{1/2}}{c^2} \tag{8.6}$$

$$= -4.442807633 \times 10^{-10}\,\mathrm{s/m}^{1/2} , \tag{8.7}$$

   where c is the speed of light and μ is the Earth's universal gravitational parameter.
4. Ionospheric correction. The SPS user can correct the time received from the satellite for ionospheric effect T_{iono} by utilizing parameters contained in page 18 of subframe 4, according to the algorithm given in [119]. It uses data transmitted by the satellite (the four coefficients of a cubic equation representing the amplitude of the vertical delay and the four coefficients of a cubic equation representing the period of the model) and some terms generated by the receiver (the elevation angle between the user and satellite, the azimuth angle between the user and satellite,

the receiver's geodetic latitude and longitude WGS-84 and the receiver's computed system time.
Application of this algorithm leads to a reduction of 50% only of the error due to the ionospheric delays.
5. Tropospheric correction T_{tropo}. The troposphere is the lower part of the atmosphere extending from ground level to from 8 to 13 km, according to the latitude. It experiences changes in temperature, pressure and humidity associated with weather changes. Precise models of tropospheric delay require estimates or measurements of these parameters and no information is transmitted by the satellite. The apparent delay introduced by the troposphere can vary between 2 m, for a satellite elevation of 90° and 30 m for an elevation of 5°.

If t'_{r} is the raw value of the time of reception of the signal and t'_{t} the raw value of the time of its transmission, the corrected time of reception of the signal is consequently,

$$t_{\text{r}} = t'_{\text{r}} - T_{\text{iono}} - T_{\text{tropo}} , \tag{8.8}$$

while the corrected time of emission of the same signal is

$$t_{\text{e}} = t'_{\text{e}} + a_{\text{f0}} + a_{\text{f1}}(t - t_{\text{oc}}) + a_{\text{f2}}(t - t_{\text{oc}})^2 - T_{\text{GD}} + \Delta t_{\text{r}} \tag{8.9}$$

and the pseudo-range PR, corrected from these effects, is

$$PR = c(t_{\text{r}} - t_{\text{e}}) . \tag{8.10}$$

Moreover, the clock of the receiver is not exactly set to the GPS time and must be corrected from an error δt. This correction is one of the four unknown quantities calculated by the receiver from the four (or more) pseudo-range measurements.

Determination of Coordinates and Time

The calculations are performed in the ECEF[4] reference system.

The value of at least four corrected ranges CR_i, $1 \leq i \leq 4$ between the receiver and at least four satellites S_i allows the calculation of the three spatial coordinates X, Y, Z of the receiver in this reference system and the time correction δt to apply to the receiver clock. X, Y, Z and δt are the four unknown quantities of this calculation. The distance ρ_i between the receiver and the clock C_i is

[4] ECEF: Earth centered, Earth fixed. This is the reference system whose origin is at the Earth's center, whose X-axis crosses the equator at the prime meridian Z-axis is the axis of rotation of the earth and whose Y-axis forms a right-handed system with the X-axis and the Z-axis. The distances in this system are expressed in meters. The conversion in latitude, longitude and altitude relies on the definition of the geoid ellipsoid.

$$\rho_i = \sqrt{(X_i - X)^2 + (Y_i - Y)^2 + (Z_i - Z)^2}\ . \tag{8.11}$$

The spatial coordinates X_i, Y_i and Z_i of the satellite S_i are given by the navigation message. It is to be noted that the navigation message gives the ephemeris data in the WGS 84[5] reference system. The distance ρ_i can be expressed as

$$\rho_i == CR_i + c\delta t\ . \tag{8.12}$$

The quantity $c\delta t$ is the correction related to the error of the receiver time. The four unknown quantities X, Y, Z and δt can consequently be deduced from the four equations

$$\begin{cases} \sqrt{(X_1 - X)^2 + (Y_1 - Y)^2 + (Z_1 - Z)^2} = CR_1 + c\delta t\ , \\ \sqrt{(X_2 - X)^2 + (Y_2 - Y)^2 + (Z_2 - Z)^2} = CR_2 + c\delta t\ , \\ \sqrt{(X_3 - X)^2 + (Y_3 - Y)^2 + (Z_3 - Z)^2} = CR_3 + c\delta t\ , \\ \sqrt{(X_4 - X)^2 + (Y_4 - Y)^2 + (Z_4 - Z)^2} = CR_4 + c\delta t\ . \end{cases} \tag{8.13}$$

A simple algorithm for the calculation of X, Y, Z and δt from these four equations is given in Sect. B.7. This algorithm can be applied in the case where an estimate of the position of the receiver is known and gives the correction to this estimate.

The accuracy of the result depends on the relative position of the satellites and of the receiver. This is characterized by the geometric dilution of precision (GDOP). Figure 8.6 shows a two-dimensional model of a good relative position while Fig. 8.7 shows a poor one.

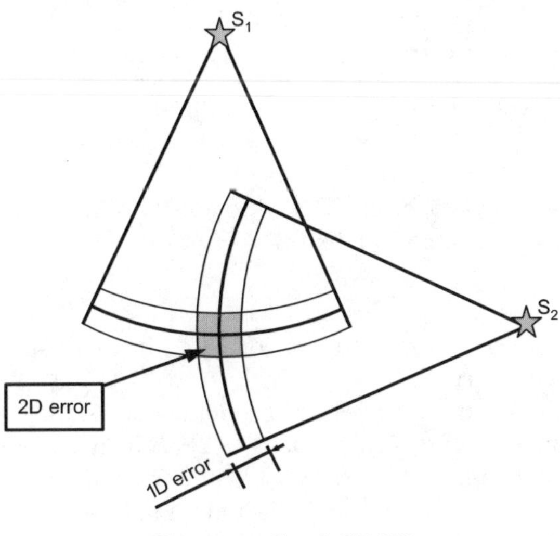

Fig. 8.6. Good GDOP

[5] WGS 84 stands for world geodetic system 1984. This reference system agrees with the international terrestrial reference frame – ITRF.

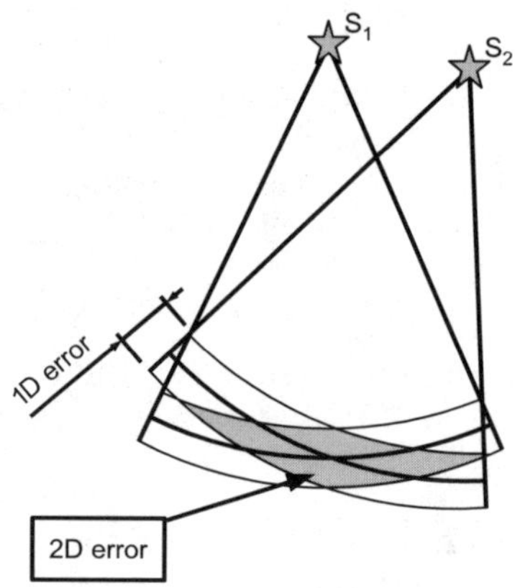

Fig. 8.7. Poor GDOP

Poor GDOP results when angles from the receiver to the set of satellites used are not very different (Fig. 8.7) while good GDOP results when angles from the receiver to satellites are quite different (Fig. 8.6).

GDOP is computed from the geometric disposition of the receiver and the satellites used by the receiver. For planning purposes GDOP is often computed from almanacs and an estimated position of the receiver (which does not need to be known with great accuracy). Of course, the estimated GDOP does not take into account obstacles that may block the line-of-sight from the receiver to the satellites. Consequently, an estimated GDOP may not be realizable in the field.

The Main Error Sources

GPS errors are a combination of noise and bias.

Noise Errors

They are the combined effect of PRN code noise (around 1 meter) and noise within the receiver (around 1 meter).

Bias Errors

1. Satellite clock errors that are not corrected by the control segment can result in 1 meter errors.
2. Ephemeris data errors can result in 1 meter errors.
3. The error on the tropospheric delays can result in 1 meter errors.
4. The error on the ionosphere delays results in 10 meter errors.
5. Multipath error are caused by reflected signals from surfaces near the receiver that can either interfere with or be mistaken for the direct signal. The resulting error is less than 1 meter.

Some much larger errors can result of an incorrect use of the receiver.

GDOP

In general, these errors must be multiplied by the appropriate GDOP term to estimate the resulting position or time error. Various GDOP terms can be computed.

- ECEF XYZ DOP terms can be rotated into a North–East down (NED) system to produce local horizontal and vertical DOP terms.
- PDOP is the (3D) position dilution of precision.
- HDOP is the horizontal dilution of precision (latitude, longitude).
- VDOP is the vertical dilution of precision (altitude).
- TDOP is the time dilution of precision (time).

8.2.3 Other Algorithms

The fundamental processing of the data transmitted by the satellites is described above. Other algorithms are possible.

Using the P-code

The principle is the same as for the C/A code, with the following improvements:

1. The bit rate is ten times the bit rate of the C/A code. The synchronization of the receiver can consequently be more accurate.
2. The length of the sequence is one week; there is no ambiguity in the pseudo-range.
3. The code modulates the two carriers L_1 and L_2 whose frequencies are different (1 575.42 and 1 227.60 MHz).
   The time difference between the two received sequences (which are emitted at the same time) is due to the ionospheric delay, which depends on the frequency ν as $1/\nu^2$.
   It is consequently possible to deduce the ionospheric time delay from this time difference and the correction for this effect is thus much more precise than the application of the model emitted by the satellites.

Differential GPS (DGPS)

GPS accuracy can be improved further to about 1 cm over short distances, using differential GPS (DGPS).

DGPS uses a network of stationary, well localized, GPS receivers and calculates the difference between their actual known position and the position as calculated by their received GPS signal. This observed error is broadcast as a local FM signal, and can be used by a user located in the vicinity of each station to correct their own calculated position, assuming that their error is the same as that of the station.

Carrier Phase Tracking

The carrier of the signal emitted by the satellites allows distances of up to 30 km to be measured with a precision of a few mm.

Starting from an epoch t_0, the phase variation $\Delta\phi_i(t - t_0)$ of the carrier of satellite S_i is continuously measured by the receiver. Assuming that the satellite clock is perfectly stable and that the ionospheric and tropospheric delays are constant during the duration of the measurement, this phase variation is only due to the variation $\Delta d_i(t - t_0)$ of the distance between the receiver and the satellite S_i. If the wavelength of the carrier is λ_c, a phase variation $\Delta\phi_i$ corresponds to a distance variation

$$\Delta d_i = \frac{\lambda_c \Delta\phi_i}{2\pi} \tag{8.14}$$

$$= \frac{c\Delta\phi_i}{2\pi\nu_c} . \tag{8.15}$$

The phase of the received signal is measured against the phase of a reference oscillator; in this case, it is the clock of the receiver. Errors can consequently be due to the clock of the satellite, the clock of the oscillator and various delays such as ionospheric delay.

Phase Fluctuations and the Time Error of a Clock

The signal emitted by a clock C is

$$x(t) = X \exp[j\phi(t)] \tag{8.16}$$

$$= X \exp[j2\pi\nu_0 t + j\delta\phi(t)] , \tag{8.17}$$

where ν_0 is the nominal value of its frequency (for instance 1 575.42 MHz for the L_1 signal) and $\delta\phi(t)$ is the value of the phase fluctuation at time t.

The value $t_C(t)$ of the time given by the clock C at time t is

$$t_C(t) = \frac{\phi(t)}{2\pi\nu_0} \tag{8.18}$$

$$= t + \frac{\delta\phi(t)}{2\pi\nu_0} . \tag{8.19}$$

The time error $\delta t_C(t)$ of the clock C is then

$$\delta t_C(t) = \frac{\delta\phi(t)}{2\pi\nu_0} . \tag{8.20}$$

This means, for instance, that a time error of 1 ns produces a phase error of 9.9 rad on the L_1 signal.

Time Error and Range Error

The error δd on the measured range due to a time error δt_{C} is

$$\delta d = c \times \delta t_{\mathrm{C}} \ . \tag{8.21}$$

This means that a time error of 1 ns produces a pseudo-range error of about 30 cm.

Notice that, since the phase of the received signal is measured against the phase of the receiver clock, the time error can arise from the satellite clock, from the receiver clock or from both.

The constraints on the time error are difficult to fulfill and differential measurement are usually performed. In fact, in order to effectively get rid of the delay fluctuations, the motion of the satellites and the satellite clock fluctuations, all carrier-phase tracking is differential and at least two receivers track carrier signals at the same time. Delay differences at the two receivers must be small enough to insure that the ionospheric and tropospheric perturbations are the same. This usually requires that the two receivers be within about 30 km of each other.

Differential Phase Measurements

Let

- $\delta\phi_{\mathrm{s}}(t)$ be the phase fluctuation of the satellite clock during the measurement;
- $\rho_1(t)$ be the distance between the satellite and receiver number 1. This receiver is fixed, its position is known with a great accuracy: it is the reference position;
- $\rho_2(t)$ be the distance between the satellite and receiver number 2;
- $\delta\tau_1(t)$ be the delay fluctuation on the path satellite $\rightarrow$ receiver number 1,
- $\delta\tau_2(t)$ be the delay fluctuation on the path satellite $\rightarrow$ receiver number 2.

The phase of the satellite clock at time t is

$$\phi_{\mathrm{s}}(t) = 2\pi\nu_{\mathrm{s}}t + \delta\phi_{\mathrm{s}}(t) \ . \tag{8.22}$$

The phase $\phi_{\mathrm{s},1}(t)$ of the emitted wave arriving on the reference receiver 1 is

$$\phi_{\mathrm{s},1}(t) = 2\pi\nu_{\mathrm{s}}\left(t - \frac{\rho_1(t)}{c} - \delta\tau_1(t)\right) + \delta\phi_{\mathrm{s}}\left(t - \frac{\rho_1}{c} - \delta\tau_1(t)\right) \ . \tag{8.23}$$

The phase $\phi_2(t)$ of the emitted wave arriving on receiver 2 is

$$\phi_2(t) = 2\pi\nu_{\mathrm{s}}\left(t - \frac{\rho_2(t)}{c} - \delta\tau_2(t)\right) + \delta\phi_{\mathrm{s}}\left(t - \frac{\rho_2}{c} - \delta\tau_2(t)\right) \ . \tag{8.24}$$

Each of these phases is measured against the phase of the receiver clock. Receiver 1 measures the phase $\phi_{m,1}$

$$\begin{aligned}\phi_{m,1}(t) = 2\pi\nu_{\mathrm{s}}\left(t - \frac{\rho_1(t)}{c} - \delta\tau_1(t)\right) + \delta\phi_{\mathrm{s}}\left(t - \frac{\rho_1}{c} - \delta\tau_1(t)\right) \\ - 2\pi\nu_1 t - \delta\phi_1(t)\ . \qquad (8.25)\end{aligned}$$

Receiver 2 measures the phase $\phi_{m,2}$

$$\begin{aligned}\phi_{m,2}(t) = 2\pi\nu_{\mathrm{s}}\left(t - \frac{\rho_2(t)}{c} - \delta\tau_2(t)\right) + \delta\phi_{\mathrm{s}}\left(t - \frac{\rho_2}{c} - \delta\tau_2(t)\right) \\ - 2\pi\nu_2 t - \delta\phi_2(t)\ . \qquad (8.26)\end{aligned}$$

The difference $\phi_{m,2}(t) - \phi_{m,1}(t)$ is

$$\begin{aligned}\phi_{m,2}(t) - \phi_{m,1}(t) = 2\pi\left[\nu_{\mathrm{s}}\left(t - \frac{\rho_2(t)}{c} - \delta\tau_2(t)\right) - \nu_{\mathrm{s}}\left(t - \frac{\rho_1(t)}{c} - \delta\tau_1(t)\right)\right] \\ + \delta\phi_{\mathrm{s}}\left(t - \frac{\rho_2}{c} - \delta\tau_2(t)\right) - \delta\phi_{\mathrm{s}}\left(t - \frac{\rho_1}{c} - \delta\tau_1(t)\right) \\ + 2\pi\nu_1 t + \delta\phi_1(t) - 2\pi\nu_2 t - \delta\phi_2(t)\ . \qquad (8.27)\end{aligned}$$

The nominal frequencies ν_{s}, ν_1, ν_2 of the oscillators are all equal. The frequency errors are included in the phase fluctuations $\delta\phi(t)$,

$$\begin{aligned}\phi_{m,2}(t) - \phi_{m,1}(t) = 2\pi\nu_{\mathrm{s}}\left(\frac{\rho_1(t) - \rho_2(t)}{c} + \delta\tau_1(t) - \delta\tau_2(t)\right) \\ + \delta\phi_{\mathrm{s}}\left(t - \frac{\rho_2}{c} - \delta\tau_2(t)\right) - \delta\phi_{\mathrm{s}}\left(t - \frac{\rho_1}{c} - \delta\tau_1(t)\right) \qquad (8.28) \\ + \delta\phi_1(t) - \delta\phi_2(t)\ . \qquad (8.29)\end{aligned}$$

The difference $\rho_1(t) - \rho_2(t)$ of the distances between the reference position and the satellite and between the measured position and the satellite is consequently

$$\begin{aligned}\rho_1(t) - \rho_2(t) = \lambda_{\mathrm{s}}(t) \times \frac{\phi_{\mathrm{m},2}(t) - \phi_{\mathrm{m},1}(t)}{2\pi} \\ + c \times (\delta\tau_2(t) - \delta\tau_1(t)) \\ + \lambda_{\mathrm{s}} \times \frac{\delta\phi_{\mathrm{s}}\left(t - \frac{\rho_1}{c} - \delta\tau_1(t)\right) - \delta\phi_{\mathrm{s}}\left(t - \frac{\rho_2}{c} - \delta\tau_2(t)\right)}{2\pi} \\ + \lambda_s \times \frac{\delta\phi_2(t) - \delta\phi_1(t)}{2\pi}\ . \qquad (8.30)\end{aligned}$$

The application of the method relies on the assumption that

– the uncertainty on the difference $\delta\tau_1(t) - \delta\tau_2(t)$ can be neglected, and

– the phase fluctuation of the emitter on the short time interval,

$$\frac{\rho_1(t) - \rho_2(t)}{c} + \delta\tau_1(t) - \delta\tau_2(t)$$

can be neglected.

Nevertheless, the phase error of the two receiver remains and must be much smaller than 2π in order to achieve centimetric accuracy. It is to be noted that no long-term stability is required for the clock of the satellite.

With these assumptions, the range difference $\rho_1(t) - \rho_2(t)$ can be monitored continuously, through the phase difference between the two receivers. This allows the algorithm of Sect. B.7 to be applied and the distance between the two receivers to be precisely measured.

The effect of the phase error of the receivers can be strongly reduced by performing the same measurement with a second satellite (second difference).

The use of both carriers L_1 and L_2 allows the ionospheric delay to be measured, since it depends of the frequency according to a known law. In this case, it is consequently possible to perform phase measurements over longer baselines (hundreds of kilometers).

Doppler Measurement

The Doppler frequency shift of the received signal is due to the relative velocity of the emitting satellite and of the receiver in the ECEF reference system. The velocity of the satellite is known from the navigation message. The residual Doppler frequency shift can consequently be used to compute the velocity of the receiver.

This computation can be combined with the value obtained by dividing the components of the difference of two successive positions by the time interval between these two positions.

8.2.4 GPS Augmentation: WAAS and LAAS

Satellite Based Augmentation Systems

WAAS (wide area augmentation system) is a system of satellites and ground stations that provides GPS signal corrections, giving better position accuracy. These corrections are broadcast by geostationary satellites and are directly integrated by most of the GPS receivers. A WAAS-capable receiver can give a position accuracy better than three meters in both the horizontal and vertical dimensions, 95 percent of the time.

WAAS has been developed for civil aviation applications. It uses a network of 25 precisely-located ground reference stations that monitor GPS satellite signals. These stations are located throughout the continental United States, Hawaii, Puerto Rico and Alaska, with additional stations being installed in

Alaska, Canada and Mexico. These stations collect and process GPS information and send this information to WAAS master stations, which then develop a WAAS correction message that is sent to user receivers via navigation transponders on geostationary satellites. These correction messages concern the orbit parameters and the ionospheric and tropospheric delays. The European geostationary navigation overlay system (EGNOS) and the Japanese multifunctional satellite augmentation system (MSAS) are intended to bring the same wide area augmentation to Europe and to Asia, respectively.

Ground Based Augmentation Systems

A ground based augmentation system provides correction to GPS receivers at a local scale (an airport, for instance, in order to provide an all-weather landing system). The local area augmentation system (LAAS) is based on real-time differential correction of the GPS signal: local reference receivers send data to a central location at the airport, which formulate a correction message transmitted to users via a VHF data link. A receiver on an aircraft uses this information to correct GPS signals. It is similar to the DGPS technique.

8.3 Other Global Systems: GLONASS and Galileo

8.3.1 GLONASS – GLObal'naya NAvigatsionnaya Sputnikovaya Sistema

Up-to-date information on GLONASS is given in [36]. The structure of GLONASS is very similar to that of GPS.

Space Segment

GLONASS relies on 24 satellites, 11 being operational (03-31-2006). The fully deployed GLONASS satellite constellation will be composed of 24 satellites in 3 orbital planes, 8 satellites being equally spaced in each plane. The planes are inclined by 64.8 degrees relative to the equator plane. Each GLONASS satellite operates in circular 19 100 km orbit and completes an orbit in approximately 11 hours 15 minutes.

Control Segment: Ground Based Control Complex

The ground based control complex (GCS) consists the System Control Center (Krasnoznamensk, Moscow region) and some command tracking stations (CTS) located in Russia. The command tracking stations accumulate ranging data and telemetry from the satellite signals. The information from the

CTS is processed at the System Control Center, which allows updating of the navigation message of each satellite. This update is transmitted to the satellites via the CTS, whose ranging data are periodically calibrated using laser ranging devices at the quantum optical tracking stations within the GCS. Each GLONASS satellite carries laser reflectors for this purpose.

The central synchronizer is the highly precise hydrogen atomic clock forming the GLONASS system time scale. The inboard time scales (on the basis of satellite cesium atomic clocks) of all the GLONASS satellites are synchronized with the state standard UTC (CIS) in Mendeleevo, Moscow region, through the GLONASS System time scale.

Emitted Data

Each GLONASS satellite transmits two types of signals: standard precision (SP) and high precision (HP). The standard precision signal L_1 has frequency division multiple access in the L-band: $L_1 = 1602\,\text{MHz} + n \times 0.5625\,\text{MHz}$, where n is the frequency channel number ($n = 0, 1, 2, \ldots$). Each satellite transmits signal on its own frequency, which differs from that of other satellites. However, some satellites have the same frequencies but they are placed in antipodal slots of orbit planes and they do not appear at the same time in a user's view.

8.3.2 Galileo

The European global navigation satellite system (Galileo) is under civilian control and interoperable with GPS and GLONASS.

Space Segment

The first experimental satellite of the joint project of the European Commission and the European Space Agency Galileo was launched on 28 December 2005; its full operational capability should be reached in 2010. It will be interoperable with GPS and GLONASS, the two other global satellite navigation systems. The fully deployed Galileo space segment will consist of 30 satellites (27 operational and 3 active spares), positioned in three circular planes (at an altitude above Earth of 23 616 km), the orbital planes being inclined at 56 degrees with reference to the equatorial plane.

The Control Segment

The control segment of Galileo will comprise the following:

1. A network of twenty Galileo sensor stations (GSS), which will monitor the satellites and provide data to the Galileo Control Centers through a redundant communications network.

2. Two Galileo control centers (GCC), which will be implemented on European ground and will use the data of the sensor stations to compute integrity information and to synchronize the time signals of all satellites and ground station clocks.
3. Five S-band up-link stations and 10 C-band up-link stations will be installed around the globe to transmit data from the control centers to the satellites.

Emitted Data

Galileo will offer dual frequencies signals as standard, consequently allowing the direct measurement of the ionospheric delay by comparing the delay for each frequency. The real-time positioning accuracy will consequently be of the order of one meter, which is unprecedented for a publicly available system.

Galileo will further guarantee the availability of the service under all but the most extreme circumstances and will inform users in near real-time of the failure of any satellite. This will make it suitable for applications where safety is crucial.

Galileo will furthermore provide a global search and rescue (SAR) function, based on the operational Cospas–Sarsat system[6].

[6] The objective of the Cospas–Sarsat system is to reduce, as far as possible, delays in the provision of distress alerts to SAR services, and the time required to locate distress and provide assistance at sea or on land.

9 Geophysics and Radio-Astronomy: VLBI – Very Long Base Interferometry

VLBI is an interferometry technique used in radio astronomy, in which two or more signals, coming from the same astronomical object, are received by antennas that are *very distant from each other*, recorded and then correlated in deferred time (Fig. 9.1). Due to the very long distance between the receivers and the fact that the resolution is proportional to that distance, a very high resolution can be obtained (see, for instance [26]).

In conventional interferometry techniques, the signals received by the antennas are directly transmitted via a physical link to the correlator, which produces the interference fringe in real-time; the antennas are physically connected to the correlator.

In VLBI, the received signals cannot be transmitted directly and in real-time to the correlator; the propagation time fluctuations in the physical links would completely cancel the correlation between them.

On the contrary, the signals are combined in differed time; they are converted to a lower standard frequency (IF) and recorded at each telescope on magnetic tape or hard disk, with a precise time base. The recorded

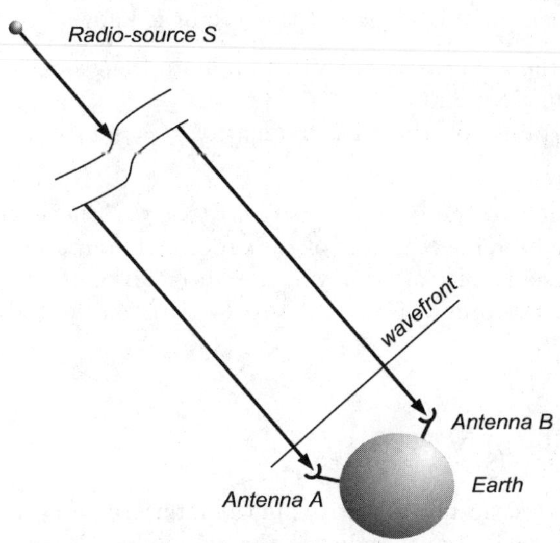

Fig. 9.1. The principle of VLBI

signals are then sent to a correlating centre, where they are synchronized, and due to the timing information, played together and combined just as if they were coming in real-time from the antennas. The correlated data can then, for instance, be turned into images using any appropriate software.

This is possible only if the phase noise of the local oscillators that down convert the signal frequency does not blur the interference fringe, and if the timestamps are accurate and stable during the duration of the experiment. In fact, only very stable atomic frequency standards can meet these requirements.

VLBI is most often performed at radio wavelengths and the following description is limited to radio signals; however, the technique has been extended to optics. The principle is very simple. Let

- **AB** be the baseline of an array of two antennas. It is the vector position of one antenna (B) with respect to the other (A).
- **s** be a unit vector in the direction of the source.

The time interval τ_{AB} (between the arrival of a wave front) to the antennas is

$$\tau_{\mathrm{AB}} = \frac{\mathbf{B} \bullet \mathbf{u}}{c}\,, \tag{9.1}$$

where c is the light velocity.

The measurement of τ_{AB} can provide one of the following types of information:

- the component of **s** along **AB** if this vector is known, or
- the component of **AB** along **s** if this vector is known.

Consequently, the applications of VLBI apply to the geodesic domain as well as the astronomic domain.

If the uncertainty on the measurement of τ_{AB} is 1 ps (1×10^{-12} s), (9.1) shows that

- if the position of the source is perfectly known, the uncertainty on the value of the baseline length is of the order of 1 mm, and
- if the baseline is perfectly known, the uncertainty on the position of the source is of the order of 1×10^{-9} rd ($\approx 10^{-3}$ arcsecond) for a baseline length of 1 000 km.

9.1 Principle of VLBI

The following description of astronomical interferometry is limited to 1D models but can easily be extended to the 2D model.

9.1.1 Interferometry

The application of interference methods to provide better resolution in astronomical measurements (in both optical and radio domains) is not a new concept (see, for instance [93, 108]).

The principle is the following: Consider two (optical or radio) receivers A and B, separated by a distance D and receiving the electromagnetic radiation emitted by a point source whose direction is at an angle α (see Fig. 9.2).

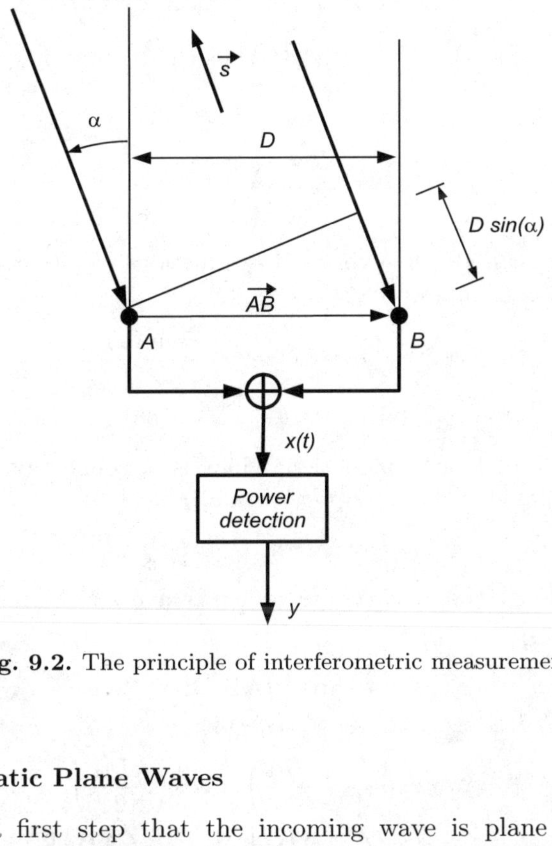

Fig. 9.2. The principle of interferometric measurements

Monochromatic Plane Waves

Consider in a first step that the incoming wave is plane and perfectly monochromatic, the frequency is ν, the wavelength is λ, the wave number is $k = \frac{2\pi}{\lambda}$ and the amplitude is X. The source is very far from Earth and its direction is indicated by the unit vector $\mathbf{s}$. The direction of the wave propagation is given by the unit vector $\mathbf{e} = -\mathbf{s}$. The equation of the wave is as follows:

$$x(t, \mathbf{r}) = X \exp \left[2\pi\nu \left(t - \frac{\mathbf{r} \cdot \mathbf{e}}{c} \right) \right] . \tag{9.2}$$

The vector $\mathbf{r} = \mathbf{OM}$ corresponds to a point M in the vicinity of the Earth. The origin O of $\mathbf{r}$ is the barycenter of the geoid, for instance. The two receivers are located at points A and B, respectively.

Supposing that the wave front is not perturbed by the atmosphere, the antennas receive the signals $x_{\mathrm{A}}(t)$ and $x_{\mathrm{B}}(t)$,

$$x_{\mathrm{A}}(t) = X \exp\left[2j\pi\nu\left(t - \frac{\mathbf{r}_{\mathrm{A}} \cdot \mathbf{e}}{c}\right)\right] \tag{9.3}$$

and

$$\begin{aligned} x_{\mathrm{B}}(t) &= X \exp\left[2j\pi\nu\left(t - \frac{\mathbf{r}_{\mathrm{B}} \cdot \mathbf{e}}{c}\right)\right] &(9.4)\\ &= x_{\mathrm{A}} \exp\left[-2j\pi\nu\left(\frac{\mathbf{AB} \cdot \mathbf{e}}{c}\right)\right] &(9.5)\\ &= x_{\mathrm{A}} \exp\left[j\left(\mathbf{AB} \cdot \mathbf{k}\right)\right] . &(9.6)\end{aligned}$$

The vector $\mathbf{k}$ is

$$\begin{aligned} \mathbf{k} &= \frac{2\pi}{\lambda}\mathbf{e} &(9.7)\\ &= k\mathbf{e} . &(9.8)\end{aligned}$$

$x_{\mathrm{A}}(t)$ and $x_{\mathrm{B}}(t)$ can also be expressed as functions of the baseline length D and the direction of the source α,

$$\begin{aligned} x_{\mathrm{B}}(t) &= x_{\mathrm{A}} \exp\left(-2j\pi\frac{D\sin(\alpha)}{\lambda}\right) &(9.9)\\ &= x_{\mathrm{A}} \exp\left(-jkD\sin(\alpha)\right) . &(9.10)\end{aligned}$$

Using the small angle approximation, which is of course not necessary (the source position angle α being supposed small),

$$x_{\mathrm{B}}(t) = x_{\mathrm{A}} \exp\left(-jkD\alpha\right) . \tag{9.11}$$

The two signals $x_{\mathrm{A}}(t)$ and $x_{\mathrm{B}}(t)$ are added to give $x(t)$,

$$\begin{aligned} x(t) &= x_{\mathrm{A}}(t) + x_{\mathrm{B}}(t) &(9.12)\\ &= x_{\mathrm{A}} \times \left[1 + \exp\left(-j\mathbf{AB} \cdot \mathbf{k}\right)\right] &(9.13)\\ &= x_{\mathrm{A}} \times \left[1 + \exp\left(-jkD\alpha\right)\right] &(9.14)\\ &= 2x_{\mathrm{A}} \exp\left(-j\frac{kD\alpha}{2}\right) \times \cos\left(\frac{kD\alpha}{2}\right) &(9.15)\\ &= 2x_{\mathrm{A}} \exp\left(-j\frac{\mathbf{AB} \cdot \mathbf{k}}{2}\right) \times \cos\left(\frac{\mathbf{AB} \cdot \mathbf{k}}{2}\right) . &(9.16)\end{aligned}$$

The output of the square law detector is, consequently,

$$\begin{aligned} y(\alpha) &= 4X^2 \cos^2\left(\frac{kD\alpha}{2}\right) &(9.17)\\ &= 4X^2 \cos^2\left(\frac{\mathbf{AB} \cdot \mathbf{k}}{2}\right) &(9.18)\\ &= 2X^2[1 + \cos(\mathbf{AB} \cdot \mathbf{k})] . &(9.19)\end{aligned}$$

This is the classical interference pattern for monochromatic radiation. The central fringe is obtained for $\alpha = 0$ ($\tau_{\mathrm{AB}} = 0$).

For $\alpha \neq 0$, the value of τ_{AB} can be measured by introducing in one of the arms of the interferometer a delay $\tau_{\mathrm{AB}'} = \pm\tau_{\mathrm{AB}}$, which compensates τ_{AB}.

Due to the rotation of the Earth, the value of α (or **AB**) varies continuously and a series of configurations can consequently be studied.

Quasi-Monochromatic Plane Wave

In fact, the radiation emitted by the source is never perfectly monochromatic. Phase and amplitude fluctuations occur,

$$x(t, \mathbf{r}) = X(t) \exp\left[2j\pi\nu\left(t - \frac{\mathbf{r}\cdot\mathbf{e}}{c}\right)\right] \tag{9.20}$$

with

$$X(t) = X_0[1 + a(t)] \exp[j\phi(t)] \;, \tag{9.21}$$

where $a(t)$ represents the relative amplitude fluctuations and $\phi(t)$ the phase fluctuations. These fluctuations are small,

$$|a(t)| \ll 1 \;, \tag{9.22}$$

$$\left|\frac{\mathrm{d}\phi(t)}{\mathrm{d}t}\right| \ll 2\pi\nu \;. \tag{9.23}$$

The signals received by the two antennas are

$$x_{\mathrm{A}}(t) = X\left(t - \frac{\mathbf{r}_{\mathrm{A}}\cdot\mathbf{e}}{c}\right) \exp\left[2j\pi\nu\left(t - \frac{\mathbf{r}_{\mathrm{A}}\cdot\mathbf{e}}{c}\right)\right] \tag{9.24}$$

$$= X\left(t - \tau_{\mathrm{A}}\right) \exp\left[2j\pi\nu\left(t - \tau_{\mathrm{A}}\right)\right] \tag{9.25}$$

$$= X\left(t_{\mathrm{A}}\right) \exp\left(2j\pi\nu t_{\mathrm{A}}\right) \tag{9.26}$$

and

$$x_{\mathrm{B}}(t) = X\left(t - \frac{\mathbf{r}_{\mathrm{B}}\cdot\mathbf{e}}{c}\right) \exp\left[2j\pi\nu\left(t - \frac{\mathbf{r}_{\mathrm{B}}\cdot\mathbf{e}}{c}\right)\right] \tag{9.27}$$

$$= X\left(t_{\mathrm{A}} - \tau_{\mathrm{AB}}\right) \exp\left[2j\pi\nu\left(t_{\mathrm{A}} - \tau_{\mathrm{AB}}\right)\right] \;. \tag{9.28}$$

In these expressions,

$$\tau_{\mathrm{A}} = \frac{\mathbf{r}_{\mathrm{A}}\cdot\mathbf{e}}{c} \;,\; t_{\mathrm{A}} = t - \tau_{\mathrm{A}} \;,\; \tau_{\mathrm{B}} = \frac{\mathbf{r}_{\mathrm{B}}\cdot\mathbf{e}}{c} \;,\; \tau_{\mathrm{AB}} = \tau_{\mathrm{B}} - \tau_{\mathrm{A}} = \frac{\mathbf{AB}\cdot\mathbf{e}}{c} \;. \tag{9.29}$$

The sum of the two signals $x_{\mathrm{A}}(t)$ and $x_{\mathrm{B}}(t)$ is

$$\begin{aligned} x_{\mathrm{A}}(t)+x_{\mathrm{B}}(t) &= X(t_{\mathrm{A}})\exp\left[2j\pi\nu(t_{\mathrm{A}})\right] \\ &\quad + X(t_{\mathrm{A}}-\tau_{\mathrm{AB}})\exp\left[2j\pi\nu(t_{\mathrm{A}}-\tau_{\mathrm{AB}})\right] \qquad (9.30)\\ &= \left[X(t_{\mathrm{A}})+X(t_{\mathrm{A}}-\tau_{\mathrm{AB}})\exp(-2j\pi\nu\tau_{\mathrm{AB}})\right] \\ &\quad \times \exp(2j\pi\nu t_{\mathrm{A}}) \ . \qquad (9.31) \end{aligned}$$

The output of the square law detector is

$$\begin{aligned} |x_{\mathrm{A}}(t)+x_{\mathrm{B}}(t)|^2 &= |X(t_{\mathrm{A}})|^2+|X(t_{\mathrm{A}}-\tau_{\mathrm{AB}})|^2 \qquad (9.32)\\ &\quad + X(t_{\mathrm{A}})X^*(t_{\mathrm{A}}-\tau_{\mathrm{AB}})\exp(2j\pi\nu\tau_{\mathrm{AB}}) \qquad (9.33)\\ &\quad + X^*(t_{\mathrm{A}})X(t_{\mathrm{A}}-\tau_{\mathrm{AB}})\exp(-2j\pi\nu\tau_{\mathrm{AB}}) \ . \qquad (9.34) \end{aligned}$$

This result is integrated over a time Δt, chosen much longer than the period of the signals but much shorter than the characteristic time of variation of the direction α of the source due to the Earth's rotation. The output $y(\tau_{\mathrm{AB}})$ of the interferometer is, consequently,

$$\begin{aligned} y(\tau_{\mathrm{AB}}) &= \left\langle |X(t_{\mathrm{A}})|^2\right\rangle + \left\langle |X(t_{\mathrm{A}}-\tau_{\mathrm{AB}})|^2\right\rangle \\ &\quad + \langle X(t_{\mathrm{A}})X^*(t_{\mathrm{A}}-\tau_{\mathrm{AB}})\rangle\exp(2j\pi\nu\tau_{\mathrm{AB}}) \\ &\quad + \langle X^*(t_{\mathrm{A}})X(t_{\mathrm{A}}-\tau_{\mathrm{AB}})\rangle\exp(-2j\pi\nu\tau_{\mathrm{AB}}) \ . \end{aligned} \qquad (9.35)$$

The mean value of the amplitude is constant,

$$\left\langle |X(t_{\mathrm{A}})|^2\right\rangle = \left\langle |X(t_{\mathrm{A}}-\tau_{\mathrm{AB}})|^2\right\rangle = X_0^2 \ . \qquad (9.36)$$

The mean values of the products

$$X^*(t_{\mathrm{A}})X(t_{\mathrm{A}}-\tau_{\mathrm{AB}})$$

and

$$X(t_{\mathrm{A}})X^*(t_{\mathrm{A}}-\tau_{\mathrm{AB}})$$

are related to the autocorrelation function $\gamma_X(t)$ of $X(t)$,

$$\gamma_X(t) = \langle X^*(\tau)X(t+\tau)\rangle \ . \qquad (9.37)$$

Consequently,

$$\begin{aligned} y(\tau_{\mathrm{AB}}) &= 2X_0^2 \\ &\quad + \gamma_X(\tau_{\mathrm{AB}})\exp(2j\pi\nu\tau_{\mathrm{AB}}) \\ &\quad + \gamma_X(-\tau_{\mathrm{AB}})\exp(-2j\pi\nu\tau_{\mathrm{AB}}) \ . \end{aligned} \qquad (9.38)$$

The following property of the autocorrelation function results from its definition (the autocorrelation is a Hermitian operator):

$$\gamma_X(-t) = \gamma_X^*(t) \ . \qquad (9.39)$$

Consequently,

$$y(\tau_{\rm AB}) = 2X_0^2 + 2\Re\left\{\gamma_X(\tau_{\rm AB})\exp\left(2j\pi\nu\tau_{\rm AB}\right)\right\} , \tag{9.40}$$

where $\Re(z)$ is the real part of the complex number z.

Notice that

$$\gamma_X(\tau_{\rm AB})\exp\left(2j\pi\nu\tau_{\rm AB}\right) = \gamma_x(\tau_{\rm AB}) , \tag{9.41}$$

where $\gamma_x(t)$ is the autocorrelation function of $x(t, \mathbf{r})$.

Finally,

$$y(\tau_{\rm AB}) = 2X_0^2 + 2\Re\left\{\gamma_x(\tau_{\rm AB})\right\} . \tag{9.42}$$

The conclusions are the following:

1. The useful information is contained in the periodic part of $y(\tau_{\rm AB})$, which is the autocorrelation function $\gamma_x(\tau_{\rm AB})$ of the plane wave emitted by the point source.
2. The time delay $\tau_{\rm AB}$ that connects the baseline and the source position appears in the value of $\gamma_x(\tau_{\rm AB})$.
3. The ratio of the periodic part of $y(\tau_{\rm AB})$ to its constant one is called the complex fringe visibility. It is proportional to the autocorrelation function of the plane wave.
4. The autocorrelation function $\gamma_x(t)$ is maximal for $t = 0$. This means that the fringe visibility is reduced when the delay $\tau_{\rm AB}$ increases, this is due to the limited coherence time of the radiation, related to its linewidth.
   The autocorrelation function is linked to the spectral density of the line (the Wiener–Khinchin theorem). If $C(t)$ is the autocorrelation function of a time function $f(t)$ whose Fourier transform is $F(\nu)$, then $C(t)$ is the Fourier transform of the absolute square of $F(\nu)$, which is the spectral density of $f(t)$.
   Consequently, the order of magnitude of the coherence time of the incoming wave is given by the inverse of its linewidth. This will ultimately limit the resolution of the observation. For instance, a linewidth of 1 kHz gives an upper limit of only 1.6×10^{-4} s for $\tau_{\rm AB}$, corresponding to

   $$|\mathbf{AB} \bullet \mathbf{s}| \approx 5\,\text{km}$$

   and, for a baseline length of 5 000 km, to $\alpha = 1 \times 10^{-3}$ rd.
   In fact, since the signals are correlated in deferred time, it is possible to shift one record until the time difference is canceled and the correlation function is maximal. The shift gives the value of the time difference $\tau_{\rm AB}$.
5. The main part of the processing of the signals received by any array of radio antennas is consequently the calculation of their correlation.
   Although the signals received by different antennas come from the same source, the quantity computed is called the cross-correlation, taking into

account the fact that each signal may have been modified in a different and non-correlated way (atmospheric perturbations, additive noise of the receiver, etc.). This calculation is made by a specialized data processing system called the *correlator*.

Extended Source

If the radiation source is extended around the incoming direction $\mathbf{e_0}$, but is incoherent (different points of the source radiate independently), there is no interference between the contribution of the different points and their power contributions are simply added.

The following model is limited to a one-dimensional source. It is easy to extend the model to the real case of two-dimensional sources.

The position of each point of the source is characterized by the unit vectors $\mathbf{e}$ (from the source) or $\mathbf{s} = -\mathbf{e}$ (toward the source) or by the angle α between the perpendicular to the baseline and $\mathbf{s}$.

The sky brightness $B(\alpha)$ of the point in the direction α is proportional to the square of the mean amplitude $X_0(\alpha)$ of the incoming radiation from that direction and the total brightness is $B_\mathrm{t} = \int_\alpha B(\alpha)\mathrm{d}\alpha$.

The output y_c of the correlator is the sum of the elementary cross-correlation functions corresponding to all the points of the source,

$$\begin{aligned} y_\mathrm{c} &= \int_\alpha \gamma_x(\alpha, \tau_\alpha)\mathrm{d}\alpha \\ &= \int_\alpha \gamma_X(\alpha, \tau_\alpha)\exp(j2\pi\nu\tau_\alpha)\,\mathrm{d}\alpha\ . \end{aligned} \tag{9.43}$$

The integral is to be taken over the radio source. Every point of the source corresponds to a value of α,

$$\alpha = \alpha_0 + \delta\alpha \tag{9.44}$$

with

$$\delta\alpha \ll 1\ , \tag{9.45}$$

where α_0 corresponds to an arbitrary reference point of the source, τ_α is the value of τ_AB for the value α of the angle between the perpendicular to the baseline, and $\mathbf{s}$

$$\begin{aligned} \tau_\alpha &= -\frac{D\sin\alpha}{c} \\ &= -\frac{D\sin\alpha_0}{c} - \frac{D\cos\alpha_0}{c}\delta\alpha\ , \end{aligned} \tag{9.46}$$

$$2\pi\nu\tau_\alpha = -2\pi\frac{D}{\lambda}\sin\alpha_0 - 2\pi\frac{D}{\lambda}\cos\alpha_0\delta\alpha\ , \tag{9.47}$$

where λ is the wavelength of the radiation. $\gamma_X(\delta\alpha, \tau_{\delta\alpha})$ is the correlation function of the sky brightness in the direction $\alpha = \alpha_0 + \delta\alpha$. Consequently,

$$y_c = \int_{\delta\alpha} \gamma_X(\delta\alpha, \tau_{\delta\alpha}) \exp\left(-j2\pi\frac{D\sin\alpha_0}{\lambda}\right) \times \exp\left(-j2\pi\frac{D\cos\alpha_0}{\lambda}\delta\alpha\right) \mathrm{d}(\delta\alpha) \,. \tag{9.48}$$

This expression shows that the output of the correlator is the Fourier transform of the function of $\delta\alpha$,

$$\gamma_{X,\alpha_0}(\delta\alpha, \tau_{\delta\alpha}) = \gamma_X(\delta\alpha, \tau_{\delta\alpha}) \exp\left(-j2\pi\frac{D\sin\alpha_0}{\lambda}\right) . \tag{9.49}$$

This function is closely related to the correlation function of the sky brightness [117],

$$y_c\left(\frac{D\cos\alpha_0}{\lambda}\right) = \mathcal{F}_\alpha[\gamma_{X,\alpha_0}(\alpha, \tau_\alpha)]\left(\frac{D\cos\alpha_0}{\lambda}\right) . \tag{9.50}$$

The conclusions are the following:

1. The correlator gives the Fourier transform of the cross-correlation function of the amplitude $X(\alpha, t_{AB})$ of the signal emitted by the source. This correlation function is calculated for the delay t_{AB} between the two receivers.
2. The sky brightness can be calculated from this result if this Fourier transform is known for different sampled values of its parameter $\frac{D\cos\alpha_0}{\lambda}$, i.e. for different values of D, the distance between the two receivers involved in the calculation of the Fourier transform and/or different values of α_0. In the first case, an array of receivers is used, in the second case, the motion of the vector **AB** due to the rotation of the Earth is used.

Examples

In the following simple examples

1. $\alpha_0 \ll 1$: $\cos\alpha_0 = 1$ and $\sin\alpha_0 = 0$.
   Consequently, (9.48) simplifies to

$$y_c = \int_{\delta\alpha} \gamma_X(\delta\alpha, \tau_{\delta\alpha}) \times \exp\left(-j2\pi\frac{D}{\lambda}\delta\alpha\right) \mathrm{d}(\delta\alpha) \tag{9.51}$$

   and the output of the correlator is the Fourier transform of the cross-correlation function $\gamma_X(\delta\alpha, \tau_{\delta\alpha})$.
2. The linewidth of the radiation emitted by the source is supposed to be narrow enough so that the cross-correlation function $\gamma_X(\delta\alpha, \tau_{\delta\alpha})$ is (a monochromatic wave)

$$\gamma_X(\delta\alpha, \tau_{\delta\alpha}) = X_0^2(\delta\alpha) \,. \tag{9.52}$$

The Rectangular Sky Brightness Function

The object is centered at $-\alpha_0/2 \ll 1$ and its width is $2\Delta\alpha$,

$$X_0^2(\delta\alpha) = \begin{cases} 0 & \begin{cases} \delta\alpha < -\alpha_0/2 - \Delta\alpha \\ \delta\alpha > -\alpha_0/2 + \Delta\alpha \end{cases} , \\ X_0^2 & -\alpha_0/2 - \Delta\alpha \leq \delta\alpha \leq -\alpha_0/2 + \Delta\alpha \, . \end{cases} \tag{9.53}$$

The Fourier transform $\mathcal{F}_\alpha[\gamma_X(\alpha, \tau_\alpha)]\left(\frac{D}{\lambda}\right)$ is

$$Y\left(\frac{D}{\lambda}\right) = 2X_0^2\delta\alpha \exp\left(-j2\pi\frac{D}{\lambda} \times \frac{-\alpha_0}{2}\right) \operatorname{sinc}\left(2\pi\frac{D}{\lambda}\delta\alpha\right) . \tag{9.54}$$

The complex fringe visibility $\Gamma(D/\lambda)$ is

$$\Gamma\left(\frac{D}{\lambda}\right) = \exp\left(-j2\pi\frac{D}{\lambda} \times \frac{-\alpha_0}{2}\right) \operatorname{sinc}\left(2\pi\frac{D}{\lambda}\delta\alpha\right) . \tag{9.55}$$

The modulus of the complex visibility is consequently maximal for small values of the ratio

$$\frac{\delta\alpha}{\lambda/D} ,$$

i.e. for objects whose angular diameter is of the order of or smaller than λ/D; VLBI is used to observe very compact sources.

A Pair of Rectangular Sky Brightness Functions

As a second example, consider a pair of rectangular sky brightness functions centered at $\pm\alpha_0/2$ and having width $2\Delta\alpha$,

$$X_0^2(\delta\alpha) = \begin{cases} 0 & \begin{cases} \delta\alpha < -\alpha_0/2 - \Delta\alpha \\ -\alpha_0/2 + \Delta\alpha < \delta\alpha < +\alpha_0/2 - \Delta\alpha \\ \delta\alpha > +\alpha_0/2 + \Delta\alpha \end{cases} , \\ X_0^2 & \begin{cases} -\alpha_0/2 - \Delta\alpha \leq \delta\alpha \leq -\alpha_0/2 + \Delta\alpha \\ +\alpha_0/2 - \Delta\alpha \leq \delta\alpha \leq +\alpha_0/2 + \Delta\alpha \, . \end{cases} \end{cases} \tag{9.56}$$

In this case, the Fourier transform $Y_2\left(\frac{D}{\lambda}\right)$ of the sky brightness function is

$$Y_2\left(\frac{D}{\lambda}\right) = 2X_0^2\delta\alpha \operatorname{sinc}\left(2\pi\frac{D}{\lambda}\delta\alpha\right) \times \left[\exp\left(-j2\pi\frac{D}{\lambda} \times \frac{-\alpha_0}{2}\right) + \exp\left(-j2\pi\frac{D}{\lambda} \times \frac{+\alpha_0}{2}\right)\right] \tag{9.57}$$

$$= 4X_0^2\delta\alpha \operatorname{sinc}\left(2\pi\frac{D}{\lambda}\delta\alpha\right) \cos\left(2\pi\frac{D}{\lambda} \times \frac{\alpha_0}{2}\right) . \tag{9.58}$$

The complex fringe visibility $\Gamma\left(\frac{D}{\lambda}\right)$ is

$$\Gamma\left(\frac{D}{\lambda}\right) = \operatorname{sinc}\left(2\pi\frac{D}{\lambda}\delta\alpha\right)\cos\left(2\pi\frac{D}{\lambda}\times\frac{\alpha_0}{2}\right) . \tag{9.59}$$

The conclusions are the following:

1. As in the previous example, the modulus of the complex visibility is maximal for small values of the ratio

$$\frac{\delta\alpha}{\lambda/D} .$$

2. The modulus of the complex visibility is maximal for small values of the quantity $(2\pi D\alpha_0)/(2\lambda)$, i.e. for

$$\frac{\lambda}{D} < \pi\alpha_0 . \tag{9.60}$$

The resolution of the interferometer is consequently given by the ratio of the distance between the two receivers to the wavelength of the radiation.

9.1.2 Processing of the Signals

The previous discussions show that processing the signal received by the antennas allows one to

- produce an image of an astronomical object (aperture synthesis);
- precisely determine the relative position of the antennas if the emitting object is distant and stable (geodesy);
- precisely determine the position of a ground or space radio source if the positions of the antennas are known;
- determine the spectra of the radio emission.

Processing at Each Antenna

The data received by the antennas are processed in the following way before being correlated (many steps of the process, such as amplification, filtering, etc. are omitted in this schematic description).

1. They are down converted to a baseband signal by mixing them with a local oscillator. The accuracy and stability of this local oscillator must be consistent with the phase shifts to be measured.
Suppose we have an input signal

$$x(t) = X\exp\left[j(2\pi\nu t + \phi)\right.$$

and a local oscillator

$$x_{\mathrm{LO}} = X_{\mathrm{LO}} \exp\left[j(2\pi\nu_{\mathrm{LO}}t + \phi_{\mathrm{LO}})\right] .$$

Mixing these two signals uses a non-linear operator, which produces output components at various frequencies, the sum and difference of the multiples of the frequencies ν and ν_{LO}. From these components, it is easy to select, with a filter, the one whose frequency is $\nu_{\mathrm{IF}} = \nu - \nu_{\mathrm{OL}}$ (this frequency is called the intermediate frequency) and whose phase is $\phi - \phi_{\mathrm{LO}}$ (these relations apply in the case where $\nu_{\mathrm{LO}} < \nu$). The baseband is centered on this frequency $\nu - \nu_{\mathrm{LO}}$. The phase fluctuations of the IF signal are consequently the sum of that of the signal and of the local oscillator,

$$x_{\mathrm{IF}} = KXX_{\mathrm{LO}} \exp j([2\pi(\nu - \nu_{\mathrm{LO}})t + \phi - \phi_{\mathrm{LO}}] . \tag{9.61}$$

This is not a problem if all the signals of the interferometer are down converted using the same local oscillator, since it is the phase difference between them that is the pertinent information. On the contrary, in the case of a VLBI, the signals from different antennas are down converted using a different local oscillator, located in the same station as the antenna; the phase of each local oscillator must consequently be very precisely defined.

2. The resulting signal is sampled and recorded in a digital media, along with a precise timestamp.
3. The recorded data are then sent to the correlator to be further processed.

Delay Compensation

Due to the delay between the two antennas whose signal are to be correlated and to the finite linewidth of the line being studied, the fringe visibility is decreased (see Sect. 9.1.1). This can be compensated, since it is possible to shift the two recorded data to optimize the value of their cross-correlation.

Digital Correlator

The correlator is the masterpiece of VLBI signal processing. Extensive descriptions can be found, for instance, in [27, 107].

In the case of a digital processing, the cross-correlation of the discrete-time process function of $f(n)$ and $g(n)$ is easily computed,

$$\gamma(n) = f \star g(n) = \sum_{p=-\infty}^{\infty} f^*(p)g(n+p) . \tag{9.62}$$

A schematic block diagram of a cross-correlator is shown in Fig. 9.3. It uses memory to implement delays of a multiple of the sampling time T_{s}, multipliers and accumulators.

In fact, the summation does not extend from $-\infty$ to $+\infty$ and the output of the device is an estimator of the cross-correlation.

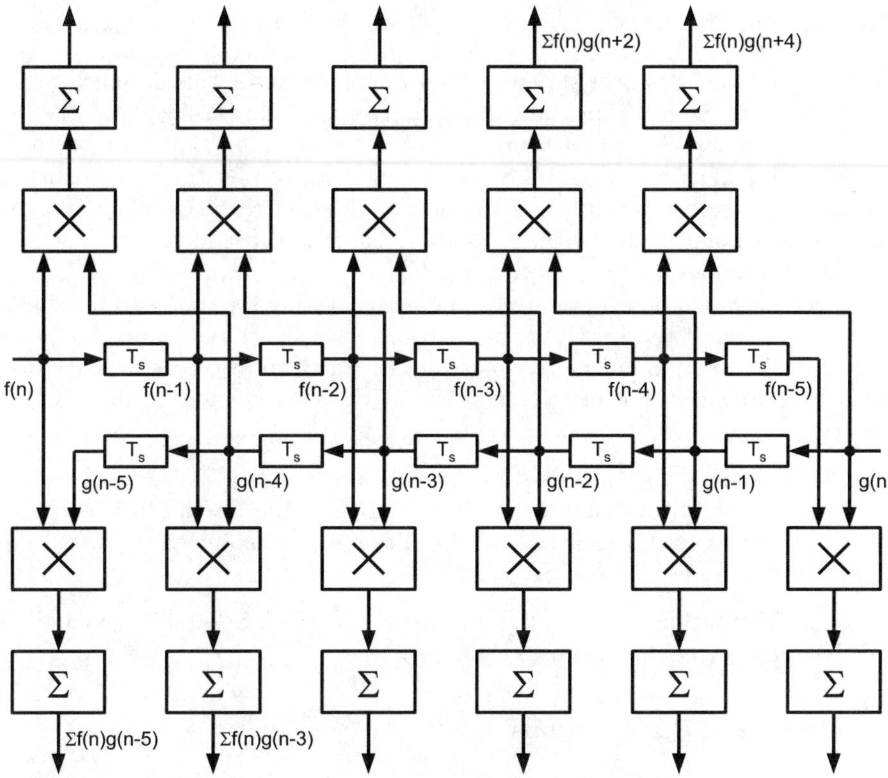

Fig. 9.3. Schematic block diagram of a cross-correlator

9.2 Applications of VLBI

It was shown in Sect. 9.1 that VLBI may have various applications in astronomy (position, spectra and imaging of astronomical objects) and in geodesy (the relative position of the antennas, absolute position relative to reference astronomical objects, rotation of the Earth).

9.2.1 Astronomy

VLBI was developed first as a radio-astronomical tool and remains a powerful and high resolution tool for observing radio sources. It allows submilliarcsecond imaging [73, 60] and detection [118] of extragalactic objects.

9.2.2 Geodesy

In this kind of application, the astronomical sources are known and used as references to determine some parameters of the Earth.

Rotation of the Earth

Very distant quasars provide an inertial reference frame that is much more accurate than the fundamental catalog of fix stars FK5 [126].

The antennas of a VLBI array are then in a situation that may be compared to that of a differential GPS experiment; they receive the signal emitted by the same source. Nevertheless, in the case of VLBI, the astronomic sources appear as a point-source with no motion. There is consequently no need to construct a model for their motion.

Since the radio telescopes are fixed on the rotating Earth, VLBI measures the orientation of the Earth in the inertial reference frame defined by these quasars as a function of time, monitoring the Earth rotation and orientation. It is consequently possible to measure all the components of the Earth's rotation:

- the position of the Earth's spin axis in space,
- the position of the Earth's spin axis relative to the Earth crust, and
- the velocity of the rotation, which allows one to connect the two time scales UT and UTC (see Sect. 7.1).

This information allow one to perform orbit controls of satellites, including GPS satellites (see, for instance [110, 95]).

Monitoring of Plate Potions

This application of VLBI, joined to the GPS technique, is well known. These space geodetic techniques allow the direct measurements of plate motions. Motions of a few cm per year are clearly visible (see, for instance [49, 7, 47]). The results of these measurements are used in Earthquake research.

Precise Localization on the Earth

The precise measurement of the position of the VLBI and GPS stations allow one to maintain the realization of the International Terrestrial Reference System (see, for instance [59, 88]).

Part IV

Appendix

A Useful Integrals

The following results are taken from the Wolfram integrator.
http://integrals.wolfram.com/

A.1 Calculation of Variances

$$\int \sin^2(x)\mathrm{d}x = \frac{x}{2} - \frac{\sin(2x)}{4} \,, \tag{A.1}$$

$$\int \frac{\sin^2(x)}{x}\mathrm{d}x = \frac{\ln(x)}{2} - \frac{\mathrm{Ci}(2x)}{2} \,, \tag{A.2}$$

$$\int \frac{\sin^2(x)}{x^2}\mathrm{d}x = \frac{\cos(2x)}{2x} + \mathrm{Si}(2x) - \frac{1}{2x} \,, \tag{A.3}$$

$$\int \frac{\sin^2(x)}{x^3}\mathrm{d}x = \frac{\cos(2x)-1}{4x^2} + \mathrm{Ci}(2x) - \frac{\sin(2x)}{2x} \,, \tag{A.4}$$

$$\begin{aligned}\int \frac{\sin^2(x)}{x^4}\mathrm{d}x &= \frac{\cos(2x)-1}{6x^3} - \frac{\cos(2x)}{3x} \\ &\quad - \frac{\sin(2x)}{6x^2} - \frac{2\mathrm{Si}(2x)}{3} \,,\end{aligned} \tag{A.5}$$

where we have the sine integral $\mathrm{Si}(x)$

$$\mathrm{Si}(x) = \int_0^x \frac{\sin(t)}{t}\mathrm{d}t \,, \tag{A.6}$$

$$\mathrm{Si}(\infty) = \pi/2 \tag{A.7}$$

and the cosine integral $\mathrm{Ci}(x)$

$$\mathrm{Ci}(x) = -\int_x^\infty \frac{\cos(t)}{t}\mathrm{d}t \tag{A.8}$$

$$= \gamma + \ln(x) + \int_0^x \frac{\cos t - 1}{t}\mathrm{d}t \,, \tag{A.9}$$

$$\mathrm{Ci}(\infty) = 0 \,. \tag{A.10}$$

For $x \ll 1$, we have

$$\mathrm{Ci}(x) \approx \gamma + \ln(x) - \frac{x^2}{4} \tag{A.11}$$

$$\approx \gamma + \ln(x)\,, \tag{A.12}$$

where γ is the Euler–Mascheroni constant

$$\gamma = 0.577216\ldots \tag{A.13}$$

A.2 Calculation of Allan Variances

$$\int \frac{\sin^4(x)}{x^2}\mathrm{d}x = \frac{4\cos(2x) - \cos(4x) - 3}{8x} + \frac{2\mathrm{Si}(2x) - \mathrm{Si}(4x)}{2x}\,, \tag{A.14}$$

$$\int \frac{\sin^4(x)}{x}\mathrm{d}x = \frac{\mathrm{Ci}(4x) - 4\mathrm{Ci}(2x) + 3\ln(x)}{8}\,, \tag{A.15}$$

$$\int \sin^4(x)\mathrm{d}x = \frac{12x - 8\sin(2x) + \sin(4x)}{32}\,, \tag{A.16}$$

$$\int \frac{\sin^4(x)}{x^3}\mathrm{d}x = \frac{4\cos(2x) - \cos(4x) - 3}{16x^2} + \mathrm{Ci}(2x) - \mathrm{Ci}(4x) + \frac{\sin(4x) - 2\sin(2x)}{4x}\,, \tag{A.17}$$

$$\int \frac{\sin^4(x)}{x^4}\mathrm{d}x = \frac{4(1-2x^2)\cos(2x) + (8x^2-1)\cos(4x) - 3}{24x^3} + \frac{\sin(4x) - 2\sin(2x)}{12x^2} + \frac{4\mathrm{Si}(4x) - 2\mathrm{Si}(2x)}{3}\,. \tag{A.18}$$

A.3 Calculation of Hadamard Variances

$$\int \frac{\sin^2(x)}{x^2}\frac{\sin^2(4x)}{\cos^2(x)}\mathrm{d}x = \frac{1}{2x}[12\cos(2x) - 8\cos(4x) + 4\cos(6x) - \cos(8x) - 7] + 12\mathrm{Si}(2x) - 16\mathrm{Si}(4x) + 12\mathrm{Si}(6x) - 4\mathrm{Si}(8x)\,, \tag{A.19}$$

$$\int \frac{\sin^2(x)}{x^2}\frac{\sin^2(6x)}{\cos^2(x)}\mathrm{d}x = \frac{1}{2x}[20\cos(2x) - 16\cos(4x) + 12\cos(6x) - 8\cos(8x) + 4\cos(10x) - \cos(12x) - 11] + 20\mathrm{Si}(2x) - 32\mathrm{Si}(4x) + 36\mathrm{Si}(6x) - 32\mathrm{Si}(8x) + 20\mathrm{Si}(10x) - 6\mathrm{Si}(12x)\,, \tag{A.20}$$

$$\int \frac{\sin^2(x)}{x}\frac{\sin^2(4x)}{\cos^2(x)}\mathrm{d}x = \frac{1}{2}\left[7\ln(x) - 12\mathrm{Ci}(2x) + 8\mathrm{Ci}(4x)\right.$$
$$\left. -4\mathrm{Ci}(6x) + \mathrm{Ci}(8x)\right] ; \tag{A.21}$$

$$\int \frac{\sin^2(x)}{x}\frac{\sin^2(6x)}{\cos^2(x)}\mathrm{d}x = \frac{1}{2}\left[11\ln(x) - 20\mathrm{Ci}(2x) + 16\mathrm{Ci}(4x) - 12\mathrm{Ci}(6x)\right.$$
$$\left. +8\mathrm{Ci}(8x) - 4\mathrm{Ci}(10x) + \mathrm{Ci}(12x)\right] , \tag{A.22}$$

$$\int \sin^2(x)\frac{\sin^2(4x)}{\cos^2(x)}\mathrm{d}x = \frac{7}{2}x - 3\sin(2x) + \sin(4x)$$
$$- \frac{1}{3}\sin(6x) + \frac{1}{16}\sin(8x) , \tag{A.23}$$

$$\int_{-\infty}^{\infty} \sin^2(x)\frac{\sin^2(6x)}{\cos^2(x)}\mathrm{d}x = \frac{11}{2}x - 5\sin(2x) + 2\sin(4x) - \sin(6x)$$
$$+ \frac{1}{2}\sin(8x) - \frac{1}{5}\sin(10x) + \frac{1}{24}\sin(12x) , \tag{A.24}$$

$$\int \frac{\sin^2(x)}{x^3}\frac{\sin^2(4x)}{\cos^2(x)}\mathrm{d}x = \frac{1}{4x^2}\left[12\cos(2x) - 8\cos(4x) + 4\cos(6x)\right.$$
$$\left. - \cos(8x) - 7\right] + 12\mathrm{Ci}(2x) - 32\mathrm{Ci}(4x)$$
$$+ 36\mathrm{Ci}(6x) - 16\mathrm{Ci}(8x) + \frac{2}{x}\left[-3\sin(2x)\right.$$
$$\left. +4\sin(4x) - 3\sin(6x) + \sin(8x)\right] , \tag{A.25}$$

$$\int \frac{\sin^2(x)}{x^3}\frac{\sin^2(6x)}{\cos^2(x)}\mathrm{d}x = \frac{1}{4x^2}\left[20\cos(2x) - 16\cos(4x) + 12\cos(6x)\right.$$
$$\left. -8\cos(8x) + 4\cos(10x) - \cos(12x) - 11\right]$$
$$+ 20\mathrm{Ci}(2x) - 64\mathrm{Ci}(4x) + 108\mathrm{Ci}(6x)$$
$$- 128\mathrm{Ci}(8x) + 100\mathrm{Ci}(10x) - 36\mathrm{Ci}(12x)$$
$$+ \frac{1}{x}\left[-10\sin(2x) + 16\sin(4x) - 18\sin(6x)\right.$$
$$\left. +16\sin(8x) - 10\sin(10x) + 3\sin(12x)\right] , \tag{A.26}$$

$$\int \frac{\sin^2(x)}{x^4}\frac{\sin^2(4x)}{\cos^2(x)}\mathrm{d}x = \frac{1}{6x^3}\left[12\cos(2x) - 8\cos(4x)\right.$$
$$\left. + 4\cos(6x) - \cos(8x) - 7\right]$$
$$+ \frac{1}{3x}\left[-12\cos(2x) + 32\cos(4x)\right.$$
$$\left. - 36\cos(6x) + 16\cos(8x)\right]$$
$$+ \frac{1}{6}\left[-48\mathrm{Si}(2x) + 256\mathrm{Si}(4x)\right.$$
$$\left. - 432\mathrm{Si}(6x) + 256\mathrm{Si}(8x)\right]$$
$$+ \frac{1}{3x^2}\left[-6\sin(2x) + 8\sin(4x)\right.$$
$$\left. -6\sin(6x) + 2\sin(8x)\right] , \tag{A.27}$$

$$
\begin{aligned}
\int \frac{\sin^2(x)}{x^4}\frac{\sin^2(6x)}{\cos^2(x)}\mathrm{d}x = {} & \frac{1}{6x^3}\left[20\cos(2x) - 16\cos(4x) + 12\cos(6x)\right. \\
& \left. -8\cos(8x) + 4\cos(10x) - \cos(12x) - 11\right] \\
& \frac{1}{3x}\left[-20\cos(2x) + 64\cos(4x) - 108\cos(6x)\right. \\
& \left. +128\cos(8x) - 100\cos(10x) + 36\cos(12x)\right] \\
& + \frac{1}{6}\left[-80\mathrm{Si}(2x) + 512\mathrm{Si}(4x) - 1\,296\mathrm{Si}(6x)\right. \\
& \left. +2\,048\mathrm{Si}(8x) - 2\,000\mathrm{Si}(10x) + 864\mathrm{Si}(12x)\right] \\
& + \frac{1}{3x^2}\left[-10\sin(2x) + 16\sin(4x) - 18\sin(6x)\right. \\
& \left. +16\sin(8x) - 10\sin(10x) + 3\sin(12x)\right] \ . \qquad \text{(A.28)}
\end{aligned}
$$

A.4 Calculation of the Three-samples Variance

$$
\int \sin(x)^6 \mathrm{d}x = \frac{5x}{16} - \frac{15}{64}\sin(2x) + \frac{3}{64}\sin(4x) - \frac{1}{192}\sin(6x)\ , \qquad \text{(A.29)}
$$

$$
\int \frac{\sin(x)^6}{x}\mathrm{d}x = \frac{5}{16}\ln(x) - \frac{15}{32}\mathrm{Ci}(2x) + \frac{6}{32}\mathrm{Ci}(4x) - \frac{1}{32}\mathrm{Ci}(6x)\ , \qquad \text{(A.30)}
$$

$$
\begin{aligned}
\int \frac{\sin(x)^6}{x^2}\mathrm{d}x = {} & \frac{1}{32x}\left[15\cos(2x) - 6\cos(4x) + \cos(6x) - 10\right] \\
& + \frac{1}{16}\left[15\mathrm{Si}(2x) - 12\mathrm{Si}(4x) + 3\mathrm{Si}(6x)\right]\ , \qquad \text{(A.31)}
\end{aligned}
$$

$$
\begin{aligned}
\int \frac{\sin(x)^6}{x^3}\mathrm{d}x = {} & \frac{1}{64x^2}\left[15\cos(2x) - 6\cos(4x) + \cos(6x) - 10\right] \\
& + \frac{1}{32}\left[30\mathrm{Ci}(2x) - 48\mathrm{Ci}(4x) + 18\mathrm{Ci}(6x)\right] \\
& + \frac{1}{32x}\left[-15\sin(2x) + 12\sin(4x) - 3\sin(6x)\right]\ , \qquad \text{(A.32)}
\end{aligned}
$$

$$
\begin{aligned}
\int \frac{\sin(x)^6}{x^4}\mathrm{d}x = {} & \frac{1}{96x^3}\left[15\cos(2x) - 6\cos(4x) + \cos(6x) - 10\right] \\
& + \frac{1}{96x}\left[-30\cos(2x) + 48\cos(4x) - 18\cos(6x)\right] \\
& + \frac{1}{48}\left[-30\mathrm{Si}(2x) + 96\mathrm{Si}(4x) - 54\mathrm{Si}(6x)\right] \\
& + \frac{1}{32x^2}\left[-5\sin(2x) + 4\sin(4x) - \sin(6x)\right]\ . \qquad \text{(A.33)}
\end{aligned}
$$

B Some Calculational Details

B.1 The Allan Filter

The Allan filter squared amplitude response is

$$|H_{\text{Allan}}(\nu)|^2 = \sin^2 \pi\nu T \ . \tag{B.1}$$

Putting $x = \pi\nu T = \pi/2 + \delta x$, we have

$$\begin{aligned} |H_{\text{Allan}}(\nu)|^2 &= \sin^2 x & \text{(B.2)} \\ &= \cos^2(\delta x) \ . & \text{(B.3)} \end{aligned}$$

Its half width $\delta\nu_{1/2} = \delta x_{1/2}/(\pi T)$ is given by

$$\cos(\delta x_{1/2}) = \frac{1}{\sqrt{2}} \ , \tag{B.4}$$

which gives

$$\begin{aligned} \delta x_{1/2} &= \frac{\pi}{4} \ , & \text{(B.5)} \\ \delta\nu_{1/2} &= 0.25 \times \nu_{\text{s}} \ . & \text{(B.6)} \end{aligned}$$

B.2 The Hadamard Filter

The squared amplitude response is (see (6.217))

$$|H_H(\nu)|^2 = \left[\frac{\sin(2M\pi\nu T)}{\cos(\pi\nu T)}\right]^2 \ . \tag{B.7}$$

$1/|\cos(\pi\nu T)|$ becomes very large when $\pi\nu T$ is close to $\pi/2 + k\pi$.

Putting $x = \pi\nu T = \pi/2 + \delta x$, we have

$$\begin{aligned} \cos(x) &= -\sin(\delta x) \ , & \text{(B.8)} \\ \sin(2Mx) &= \sin(2M\pi/2 + 2M\delta x) \\ &= \sin(M\pi) \times \cos(2M\delta x) + \cos(M\pi) \times \sin(2M\delta x) \\ &= (-1)^M \sin(2M\delta x) \ . & \text{(B.9)} \end{aligned}$$

Consequently,

$$\frac{\sin(2Mx)}{\cos(x)} = (-1)^{M+1}\frac{\sin(2M\delta x)}{\sin(\delta x)} \tag{B.10}$$

and

$$|H_H(\nu)|^2 = \left[\frac{\sin(2M\delta x)}{\sin(\delta x)}\right]^2 . \tag{B.11}$$

For $\delta x = 0$ ($\nu = \nu_s/2$)

$$|H_H(\nu_s/2)|^2 = 4M^2 . \tag{B.12}$$

This is the maximal value of the squared amplitude response of the Hadamard filter. The half power width is given by

$$\left|H_H(\nu_{1/2})\right|^2 = 2M^2 , \tag{B.13}$$

$$\left|\frac{\sin(2M\delta x_{1/2})}{\sin(\delta x_{1/2})}\right|^2 = 2M^2 . \tag{B.14}$$

Limiting the series expansion of $\sin(2M\delta x_{1/2})$ and $\sin(\delta x_{1/2})$ to the two first terms, we have

$$\sin(2M\delta x_{1/2}) \approx 2M\delta x_{1/2}\left[1 - \frac{(2M\delta x_{1/2})^2}{6}\right] , \tag{B.15}$$

$$\sin(\delta x_{1/2}) \approx \delta x_{1/2}\left[1 - \frac{\delta x_{1/2}^2}{6}\right] \tag{B.16}$$

and

$$\left|H_H(\nu_{1/2})\right|^2 \approx 4M^2\left[\frac{6 - (2M\delta x_{1/2})^2}{6 - \delta x_{1/2}^2}\right]^2 . \tag{B.17}$$

Consequently, $\delta x_{1/2}$ is given by

$$\left[\frac{6 - (2M\delta x_{1/2})^2}{6 - \delta x_{1/2}^2}\right]^2 = \frac{1}{2} \tag{B.18}$$

and

$$\delta x_{1/2} = \pm\frac{1.576}{\sqrt{5.657M^2 - 1}} \tag{B.19}$$

$$\approx \pm\frac{0.664}{M} . \tag{B.20}$$

Consequently,

$$\delta\nu_{1/2} = \frac{0.664}{\pi M} \times \nu_s \tag{B.21}$$

$$= \frac{0.211}{M} \times \nu_s \, . \tag{B.22}$$

This is an approximate value of the half width of the Hadamard filter.

B.3 Three-samples Filter

Equation (6.261) with $M = 1$ gives for its squared amplitude response

$$|H_3|^2 = \left\{1 - \frac{\cos[3\pi\nu T]}{\cos(\pi\nu T)}\right\}^2 \, . \tag{B.23}$$

Putting $x = \pi\nu T = \pi/2 + \delta x$, we have

$$\cos(x) = -\sin(\delta x) \, , \tag{B.24}$$

$$\cos(3x) = \sin(3\delta x) \, . \tag{B.25}$$

Consequently,

$$|H_3|^2 = \left[1 + \frac{\sin[3\delta x]}{\sin(\delta x)}\right]^2 \tag{B.26}$$

$$= 16\cos^4(\delta x) \, . \tag{B.27}$$

The half-width $\delta x_{1/2}$ of this filter is given by

$$\cos^4(\delta x_{1/2}) = \frac{1}{2} \, , \tag{B.28}$$

which gives

$$\delta x_{1/2} = \pm 0.572\text{rad} \, , \tag{B.29}$$

$$\delta\nu_{1/2} = \pm\frac{0.572}{\pi} \times \nu_s \tag{B.30}$$

$$= 0.182 \times \nu_s \, . \tag{B.31}$$

B.4 Phase Noise Versus Frequency Noise Spectral Density

Let the random processes $x(t)$ and $y(t)$ be defined by

$$x(t) = \frac{\delta\phi(t)}{f_0} \, , \tag{B.32}$$

$$y(t) = \frac{\delta f(t)}{f_0} \tag{B.33}$$

$$= \frac{\mathrm{d}x(t)}{\mathrm{d}t} \, . \tag{B.34}$$

The Fourier transforms are

$$X(\nu) = \int_{-\infty}^{+\infty} x(t)\exp(-2\pi j\nu t)\mathrm{d}t\ , \tag{B.35}$$

$$Y(\nu) = \int_{-\infty}^{+\infty} y(t)\exp(-2\pi j\nu t)\mathrm{d}t \tag{B.36}$$

and inversely

$$x(t) = \int_{-\infty}^{+\infty} X(\nu)\exp(+2\pi j\nu t)\mathrm{d}\nu\ , \tag{B.37}$$

$$y(x) = \int_{-\infty}^{+\infty} Y(\nu)\exp(+2\pi j\nu t)\mathrm{d}\nu\ . \tag{B.38}$$

Consequently,

$$Y(\nu) = \int_{-\infty}^{+\infty} \frac{\mathrm{d}x(t)}{\mathrm{d}t}\exp(-2j\pi\nu t)\mathrm{d}t \tag{B.39}$$

$$= \int_{-\infty}^{+\infty} \frac{\mathrm{d}}{\mathrm{d}t}\left[\int_{-\infty}^{+\infty} X(f)\exp(+2\pi jft)\mathrm{d}f\right]\exp(-2\pi j\nu t)\mathrm{d}t \tag{B.40}$$

$$= \int_{-\infty}^{+\infty}\int_{-\infty}^{+\infty} X(f)\frac{\mathrm{d}}{\mathrm{d}t}\left[\exp(+2\pi jft)\right]\mathrm{d}f\exp(-2\pi j\nu t)\mathrm{d}t \tag{B.41}$$

$$= \int_{-\infty}^{+\infty}\int_{-\infty}^{+\infty} X(f)2\pi jf\left[\exp(+2\pi jft)\right]df\exp(-2\pi j\nu t)\mathrm{d}t \tag{B.42}$$

$$= \int_{-\infty}^{+\infty}\int_{-\infty}^{+\infty} X(f)2\pi jf\times\exp\left[+2\pi j(f-\nu)t\right]\mathrm{d}f\ \mathrm{d}t\ . \tag{B.43}$$

Since

$$\int_{-\infty}^{+\infty} \exp\left[-2\pi j(\nu-f)t\right]\mathrm{d}t = \delta(\nu-f)\ , \tag{B.44}$$

we have

$$Y(\nu) = \int_{-\infty}^{+\infty} X(f)2\pi jf\times\delta(\nu-f)\times\mathrm{d}f \tag{B.45}$$

$$= 2\pi j\nu X(\nu)\ . \tag{B.46}$$

B.5 Phase Noise

The phase noise of an oscillator is frequently characterized by its value expressed in dBc as defined by the ATIS Telecom Glossary 2000 T1.523–2001 [5]: *Phase noise, $\mathcal{L}(f)$ in decibels relative to carrier power (dBc) on a 1-Hz bandwidth, is given by $\mathcal{L}(f) = 10\log[0.5(S_\phi(f))]$ where $S_\phi(f)$ is the spectral density of phase fluctuations.*

B.6 Sampling and Hold

The continuous-time function $f(t)$ is sampled and held at times nT_l. The output of the sample and hold is another continuous-time function $f'(t)$ whose Fourier transform $F'(\nu)$ is related to the Fourier transform $F(\nu)$ of the function $f(t)$.

We have

$$F'(\nu) = \int_{-\infty}^{\infty} f'(t) \times \exp(-j2\pi\nu t)\mathrm{d}t \,. \tag{B.47}$$

The continuous-time function $f'(t)$ can be written as

$$f'(t) = \sum_{n=-\infty}^{\infty} f(nT_l) \times \Pi(t - nT_l) \,. \tag{B.48}$$

In this expression, $\Pi(x)$ is a square x-function defined by

$$\Pi(x) = \begin{cases} 0 & x < 0,\ x > T \,, \\ 0.5 & x = 0,\ x > T \,, \\ 1 & 0 < x < T \,. \end{cases} \tag{B.49}$$

The Fourier transform of $f'(t)$ is

$$\begin{aligned} F'(\nu) &= \int_{n=-\infty}^{\infty} \sum_{-\infty}^{\infty} f(nT_l) \times \Pi(t - nT_l) \times \exp(-j2\pi\nu t)\mathrm{d}t \\ &= \sum_{n=-\infty}^{\infty} f(nT_l) \int_{-\infty}^{\infty} \Pi(t - nT_l) \times \exp(-j2\pi\nu t)\mathrm{d}t \,. \end{aligned} \tag{B.50}$$

However,

$$\int_{-\infty}^{\infty} \Pi(t - nT_l) \times \exp(-j2\pi\nu t)\mathrm{d}t \tag{B.51}$$

$$= \int_{nT_l}^{(n+1)T_l} \exp(-j2\pi\nu t)\mathrm{d}t \tag{B.52}$$

$$= -\frac{1}{j2\pi\nu} \{\exp[-j2\pi\nu(n+1)T_l] - \exp[-j2\pi\nu nT_l]\} \tag{B.53}$$

$$= \exp[-j2\pi\nu(n+1/2)T_l] \times T_l \times \mathrm{sinc}(\pi\nu T_l) \,. \tag{B.54}$$

Consequently,

$$\begin{aligned} F'(\nu) &= T_l \times \exp(-j\pi\nu T_l) \times \mathrm{sinc}(\pi\nu T_l) \\ &\quad \times \sum_{n=-\infty}^{\infty} f(nT_l) \exp(-j2\pi\nu nT_l) \end{aligned} \tag{B.55}$$

$$= T_l \times \exp(-j2\pi\nu(n+1/2)T_l) \times \mathrm{sinc}(\pi\nu T_l) \times F_s(\nu) \,, \tag{B.56}$$

where $F_s(\nu)$ is the Fourier transform of the sequence $\{f(nT_1)\}$,

$$F_s(\nu) = \sum_{n=-\infty}^{\infty} f(nT_1) \exp(-j2\pi\nu nT_1)dt \,. \tag{B.57}$$

On the other hand, $F_s(\nu)$ is related to the Fourier transform of the continuous-time function $f(t)$ (Nyquist–Shannon sampling theorem). Since

$$f(t) = \int_{-\infty}^{\infty} F(\nu) \exp(+j2\pi\nu t)d\nu \,, \tag{B.58}$$

the samples $f(nT_1)$ can be expressed as

$$f(nT_1) = \int_{-\infty}^{\infty} F(\nu) \exp(+j2\pi\nu nT_1)d\nu \tag{B.59}$$

$$= \sum_{k=-\infty}^{\infty} \int_{(k-1/2)/T_1}^{(k+1/2)/T_1} F(\nu) \exp(+j2\pi\nu nT_1)d\nu \tag{B.60}$$

$$= \int_{-1/2T_1}^{1/2T_1} \left[\sum_{k=-\infty}^{\infty} F(\nu' + k/T_1) \right] \exp(+j2\pi\nu' T_1)d\nu' \,, \tag{B.61}$$

with $\nu' = \nu - k/T_1$.

Consequently, the two Fourier transforms $F(\nu)$ and $F_s(\nu)$ are connected by the relation

$$F_s(\nu) = \frac{1}{T_1} \sum_{k=-\infty}^{\infty} F(\nu + k/T_1) \,. \tag{B.62}$$

Moreover, the Fourier transform $F'(\nu)$ of the sampled and held continuous-time signal is connected to the Fourier transform $F(\nu)$ of the original continuous-time signal by the relation

$$F'(\nu) = \exp(-j\pi\nu T_1) \times \mathrm{sinc}(\pi\nu T_1) \times \sum_{k=-\infty}^{\infty} F(\nu + k/T_1) \,. \tag{B.63}$$

B.7 Algorithm for a GPS Receiver

B.7.1 The Data

Estimate of the Position

An estimate of the position of the receiver is known. It is, for instance, the last position computed by the receiver, or an estimate entered by the user.

The coordinates of this estimated position are X_e, Y_e and Z_e, while the unknown coordinates of the receiver position are X, Y and Z.

Estimate of the Local Time

An estimate t_e of the time is given by the receiver clock. The time error δt must be calculated and X_e, Y_e and Z_e must be corrected.

Position of the Four Satellites

The position of each of the four satellites used by the receiver is known due to the navigation messages. Each satellite position is given by the coordinates X_i, Y_i and Z_i of the satellite number i $(1 \leq i \leq 4)$.

Distances to the Four Satellites

The time delay between the C/A-code of the receiver and the received C/A-codes – corrected for the ionospheric and tropospheric delays, the relativistic effects and the satellite clocks error – gives the pseudo-range CR_i $(1 \leq i \leq 4)$ between the receiver and each of the satellites. The ambiguity of an integer multiple of 300 km has been raised.

The distances

$$ER_i = \sqrt{(X_i - X_e)^2 + (Y_i - Y_e)^2 + (Z_i - Z_e)^2}$$

between the estimated position and the four satellites are known.

B.7.2 Algorithm

The four pseudo-ranges are the distance between the receiver and the corresponding satellite modified by the time error of the receiver clock.

Let δt be that time error such that

$$t_e = t + \delta t \; , \tag{B.64}$$

where t is the GPS time.

The pseudo-ranges are, therefore,

$$CR_i = \sqrt{(X_i - X)^2 + (Y_i - Y)^2 + (Z_i - Z)^2} - c\delta t \; , \tag{B.65}$$

where c is the light velocity.

The coordinates of the receiver can be expressed from the estimated position coordinates

$$\begin{cases} X = X_e + \delta X \; , \\ Y = Y_e + \delta Y \; , \\ Z = Z_e + \delta Z \; . \end{cases} \tag{B.66}$$

Consequently, the pseudo-ranges are

$$\begin{aligned} CR_i &= \left[(X_i - X)^2 + (Y_i - Y)^2 + (Z_i - Z)^2\right]^{1/2} - c\delta t \\ &= ER_i \\ &\quad + \left.\frac{\partial CR_i}{\partial X}\right|_{X=X_e} \delta X + \left.\frac{\partial CR_i}{\partial Y}\right|_{Y=Y_e} \delta Y + \left.\frac{\partial CR_i}{\partial Z}\right|_{Z=Z_e} \delta Z \\ &\quad - c\delta t\,. \end{aligned} \tag{B.67}$$

However,

$$\left.\frac{\partial CR_i}{\partial X}\right|_{X=X_e} = \frac{X_e - X_i}{ER_i}\,, \tag{B.68}$$

$$\left.\frac{\partial CR_i}{\partial Y}\right|_{Y=Y_e} = \frac{Y_e - X_i}{ER_i}\,, \tag{B.69}$$

$$\left.\frac{\partial CR_i}{\partial Z}\right|_{Z=Z_e} = \frac{Z_e - X_i}{ER_i}\,. \tag{B.70}$$

The following system of linear algebraic equations can then be written

$$\begin{pmatrix} \frac{X_e-X_1}{ER_1} & \frac{Y_e-X_1}{ER_1} & \frac{Z_e-X_1}{ER_1} & -1 \\ \frac{X_e-X_2}{ER_2} & \frac{Y_e-X_2}{ER_2} & \frac{Z_e-X_2}{ER_2} & -1 \\ \frac{X_e-X_3}{ER_3} & \frac{Y_e-X_3}{ER_3} & \frac{Z_e-X_3}{ER_3} & -1 \\ \frac{X_e-X_4}{ER_4} & \frac{Y_e-X_4}{ER_4} & \frac{Z_e-X_4}{ER_4} & -1 \end{pmatrix} \times \begin{pmatrix} \delta X \\ \delta Y \\ \delta Z \\ c\delta t \end{pmatrix} = \begin{pmatrix} CR_1 - ER_1 \\ CR_2 - ER_2 \\ CR_3 - ER_3 \\ CR_4 - ER_4 \end{pmatrix}, \tag{B.71}$$

or

$$A \times \delta \mathbf{M} = \delta \mathbf{R}\,. \tag{B.72}$$

If the matrix A is non-singular (if the determinant $|A| \neq 0$), the system has a unique solution in the four variables.

Cramer's rule gives

$$M_{\mathrm{k}} = \frac{|A_k|}{|A|} \tag{B.73}$$

with

$$|A| = \begin{vmatrix} \frac{X_e-X_1}{ER_1} & \frac{Y_e-X_1}{ER_1} & \frac{Z_e-X_1}{ER_1} & -1 \\ \frac{X_e-X_2}{ER_2} & \frac{Y_e-X_2}{ER_2} & \frac{Z_e-X_2}{ER_2} & -1 \\ \frac{X_e-X_3}{ER_3} & \frac{Y_e-X_3}{ER_3} & \frac{Z_e-X_3}{ER_3} & -1 \\ \frac{X_e-X_4}{ER_4} & \frac{Y_e-X_4}{ER_4} & \frac{Z_e-X_4}{ER_4} & -1 \end{vmatrix}, \tag{B.74}$$

$$|A_1| = \begin{vmatrix} CR_1 - ER_1 & \frac{Y_e - X_1}{ER_1} & \frac{Z_e - X_1}{ER_1} & -1 \\ CR_2 - ER_2 & \frac{Y_e - X_2}{ER_2} & \frac{Z_e - X_2}{ER_2} & -1 \\ CR_3 - ER_3 & \frac{Y_e - X_3}{ER_3} & \frac{Z_e - X_3}{ER_3} & -1 \\ CR_4 - ER_4 & \frac{Y_e - X_4}{ER_4} & \frac{Z_e - X_4}{ER_4} & -1 \end{vmatrix}, \tag{B.75}$$

$$|A_2| = \begin{vmatrix} \frac{X_e - X_1}{ER_1} & CR_1 - ER_1 & \frac{Z_e - X_1}{ER_1} & -1 \\ \frac{X_e - X_2}{ER_2} & CR_2 - ER_2 & \frac{Z_e - X_2}{ER_2} & -1 \\ \frac{X_e - X_3}{ER_3} & CR_3 - ER_3 & \frac{Z_e - X_3}{ER_3} & -1 \\ \frac{X_e - X_4}{ER_4} & CR_4 - ER_4 & \frac{Z_e - X_4}{ER_4} & -1 \end{vmatrix}, \tag{B.76}$$

$$|A_3| = \begin{vmatrix} \frac{X_e - X_1}{ER_1} & \frac{Y_e - X_1}{ER_1} & CR_1 - ER_1 & -1 \\ \frac{X_e - X_2}{ER_2} & \frac{Y_e - X_2}{ER_2} & CR_2 - ER_2 & -1 \\ \frac{X_e - X_3}{ER_3} & \frac{Y_e - X_3}{ER_3} & CR_3 - ER_3 & -1 \\ \frac{X_e - X_4}{ER_4} & \frac{Y_e - X_4}{ER_4} & CR_4 - ER_4 & -1 \end{vmatrix}, \tag{B.77}$$

$$|A_4| = \begin{vmatrix} \frac{X_e - X_1}{ER_1} & \frac{Y_e - X_1}{ER_1} & \frac{Z_e - X_1}{ER_1} & CR_1 - ER_1 \\ \frac{X_e - X_2}{ER_2} & \frac{Y_e - X_2}{ER_2} & \frac{Z_e - X_2}{ER_2} & CR_2 - ER_2 \\ \frac{X_e - X_3}{ER_3} & \frac{Y_e - X_3}{ER_3} & \frac{Z_e - X_3}{ER_3} & CR_3 - ER_3 \\ \frac{X_e - X_4}{ER_4} & \frac{Y_e - X_4}{ER_4} & \frac{Z_e - X_4}{ER_4} & CR_4 - ER_4 \end{vmatrix}. \tag{B.78}$$

References

1. *International Conference held at Washington for the purpose of fixing a prime meridian and a universal day. October 1884. Protocols of the Proceedings.* Gibson Bros., October 1884.
2. Agilent. *Agilent 5071A Primary Frequency Standard Unsurpassed Stability in the Lab or Field Product Overview.* Agilent Technologies.
3. D.W. Allan, D.D. Davis, M. Weiss, A. Clements, B. Guinot, M. Granveaud, K. Dorenwendt, B. Fischer, P. Hetzel, S. Aoki, M.-K. Fujimoto, L. Charron, N. Ashby. Accuracy of international time and frequency comparison via global positioning system satellites in common-view. *IEEE Transactions on Instrumentation and Measurement*, IM–34, June 1985.
4. D.W. Allan. Statistics of Atomic Frequency Standards. In: *Proc. IEEE*, volume 54, p. 221, 1966.
5. Alliance for Telecommunications Industry Solutions. ATIS Telecom Glossary 2000 t1.523-2001. http://www.atis.org/.
6. M. Arditi, T.R. Carver. Pressure, light, and temperature shifts in optical detection of 0–0 hyperfine resonance of alkali metals. *Physical Review*, 124:800–809, 1961.
7. D.F. Argus, W.R. Peltier. Postglacial rebound from space geodesy. *AGU Fall Meeting Abstracts*, pp. A18+, December 2005.
8. N. Ashby, D.W. Allan. Practical implications of relativity for a global coordinate time scale. *Radio Science*, 14:649–669, August 1979.
9. C. Audoin, G. Santarelli, A. Makdissi, A. Clairon. Properties of an oscillator slaved to a periodically interrogated atomic resonator. *IEEE Transactions on Ultrasonics, Ferroelectrics and Frequency Control*, 45(4), 1998.
10. A. Bauch, J. Achkar, S. Bize, D. Calonico, R. Dach, R. Hlavac, L. Lorini, T.E. Parker, G. Petit, D. Piester, K. Szymaniec, P. Uhrich. Comparison between frequency standards in Europe and the USA at the 10^{-15} uncertainty level. *Metrologia*, 43:109–120, February 2006.
11. A. Bauch, R. Schroder, S. Weyers. Discussion of the uncertainty budget and of long term comparison of PTB's primary frequency standards CS1, CS2 and CSF1. In: *Proceedings of the 2003 IEEE FCS and the 17th EFTF (Tampa, USA)*, pp. 191–198, May 2003.
12. R.A. Baugh. Frequency modulation analysis with the Hadamard variance. *Proc. Annual. Symposium on Frequency Control*, pp. 222–225, 1971.
13. C.W. Beer, R.A. Bernheim. Hyperfine pressure shift of ^{133}Cs atoms in noble and molecular buffer gases. *Physical Review A*, 13:1052–1057, 1976.
14. S. Bize, P. Laurent, M. Abgrall, H. Mariona, I. Maksimovica, L. Cacciapuoti, J. Grünert, C. Viana, F. Pereira dos Santos, P. Rosenbuscha, P. Lemonde,

G. Santarelli, P. Wolf, A. Clairon, A. Luitenb, M. Tobar, C. Salomon. Advances in atomic fountains. *C.R. Physique*, 5, 2004.

15. F. Bloch, A. Siegert. Magnetic resonance for nonrotating fields. *Physical Review*, 57:522–527, 1940.
16. E. Boileau, B. Picinbono. Statistical study of phase fluctuations and oscillator stability. *IEEE Transactions on Instrumentation and Measurement*, IM-25(1):66–75, March 1976.
17. Bureau International des Poids et Mesures. ftp://ftp2.bipm.fr/pub/tai/data/PFS_reports/.
18. Bureau International des Poids et Mesures. Acronyms and locations of the timing centres which maintain a local approximation of UTC, UTC(k), and/or an independent local time scale, TA(k). ftp://62.161.69.5/pub/tai/publication/acronyms.pdf.
19. Bureau International des Poids et Mesures. *The International System of Units (SI) 7th edition.* Organisation Intergouvernementale de la Convention du Mètre, 1998.
20. Bureau International des Poids et Mesures. *The International System of Units – Supplement 2000: Addenda and Corrigenda to the 7th Edition (1998).* Organisation Intergouvernementale de la Convention du Mètre, 2000.
21. Bureau International des Poids et Mesures. Rapport du directeur sur l'activité et la gestion du Bureau International des Poids et Mesures. Technical report, Bureau International des Poids et Mesures, 2004.
22. Bureau International des Poids et Mesures. Circular T 217. Technical report, Bureau International des Poids et Mesures, February 2006.
23. Bureau International des Poids et Mesures. Circular T 219. Technical report, Bureau International des Poids et Mesures, April 2006.
24. Bureau International des Poids et Mesures. Fractional frequency of EAL from primary frequency standards. Technical report, Bureau International des Poids et Mesures, January 2006.
25. Bureau International des Poids et Mesures. Weights of clocks participating in the computation of TAI. Technical report, Bureau International des Poids et Mesures, February 2006.
26. B.F. Burke, F. Graham-Smith. *An Introduction to Radio Astronomy.* Cambridge University Press, 2002.
27. B.R. Carlson, P.E. Dewdney, T.A. Burgess, R.V. Casorso, W.T. Petrachenko, W.H. Cannon. The S2 VLBI correlator: A correlator for space VLBI and geodetic signal processing. *The Publications of the Astronomical Society of the Pacific*, 111:1025–1047, August 1999.
28. T. Celano, J. Warriner, S. Francis, A. Gifford, P. Howe, R. Beckman. Two-way time transfer to airborne platforms using commercial satellite modems. In: *Proceedings of the 34th Annual Precise Time and Time Interval (PTTI) Meeting*, December 2002.
29. A. Clairon, S. Ghezali, G. Santelli, Ph. Laurent, S.N. Lea, M. Bahoura, E. Simon, S. Weyers, K. Szymaniec. Preliminary accuracy evaluation of a cesium fountain frequency standard. In: *Proc. of 5th Symposium on Frequency Standards and Metrology*, Singapore. World Scientific.
30. A. Clairon, P. Laurent, G. Santarelli, S. Ghezali, S.N. Lea, M. Bahoura. A cesium fountain frequency standard: Preliminary results. *IEEE Transactions on Instrumentation and Measurement*, 44, 1995.

31. D.S. Coco, C. Coker, S.R. Dahlke, J.R. Clynch. Variability of GPS satellite differential group delay biases. *IEEE Transactions on Aerospace Electronic Systems*, 27:931–938, November 1991.
32. Comité Consultatif pour la Définition de la Seconde (CCDS). Recommendation S15. In: *Procès-Verbaux des Séances du Comité International des Poids et Mesures, 38 (1970)*, volume 9, 1980.
33. Comité International des Poids et Mesures (CGPM). Recommendation S2. In: *Procès-Verbaux des Séances du Comité International des Poids et Mesures, 38 (1970)*, volume 38, 1970.
34. E.U. Condon, G.H. Shortley. *The Theory of Atomic Spectra.* Cambridge University Press, 1935.
35. Conférence Générale des Poids et Mesures (CGPM). Resolution 1: Si unit of time (second). In: *Comptes rendus of the 13th CGPM*, 1967–1968.
36. Coordinational Scientific Information Center. General glonass. http://www.glonass-center.ru/.
37. C.C. Costain, J.S. Boulanger, H. Daams, D.W. Hanson, R.E. Beehler, A.J. Clements, D.D. Davis, W.J. Klepczynski, L. Veenstra, J. Kaiser, B. Guinot, P. Azoubib, J. Parcelier, G. Freon, M. Brunet. Two-way time transfer via geostationary satellites NRC/NBS, NCR/USNO and NBS/USNO via Hermes and NRC/LPTF (France) via Symphonie. In: *Proceedings of the 1978 Annual Precise Time and Time Interval (PTTI) Meeting*, November 1978.
38. L.S. Cutler, C.A. Flory, R.P. Giffard, A. De Marchi. Erratum: Frequency pulling by hyperfine sigma-transitions in cesium beam atomic frequency standards [J. Appl. Phys. 69, 2780 (1991)]. *Journal of Applied Physics*, 70, August 1991.
39. L.S. Cutler, C.A. Flory, R.P. Giffard, A. De Marchi. Frequency pulling by hyperfine sigma transitions in cesium beam atomic frequency standards. *Journal of Applied Physics*, 69:2780–2792, March 1991.
40. E. de Clercq. L'étalon primaire de fréquence à jet de césium du BNM-SYRTE. *Bulletin du BNM*, 2004-2(125):5–21, 2004.
41. C. Degenhardt, H. Stoehr, C. Lisdat, G. Wilpers, H. Schnatz, B. Lipphardt, T. Nazarova, P.-E. Pottie, U. Sterr, J. Helmcke, F. Riehle. Calcium optical frequency standard with ultracold atoms: Approaching 10^{-15} relative uncertainty. *Physical Review A*, 72(6):062111, 2005.
42. G.J. Dick. Local oscillator induces instabilities in trapped ion frequency standards. In: *Proccedings of the 19th Precise Time and Time Interval (PTTI) Meeting*, 1987.
43. efunda. efunda engineering fundamentals. http://www.efunda.com/.
44. Albert Einstein. Die grundlage der allgemeinen relativitätstheorie. *Annalen der Physik*, 49, 1916.
45. C.R. Ekstrom, W.M. Golding, R.E. Drullinger, F.L. Walls, A. DeMarchi, S.L. Rolston, W. Phillips. The Design of an Atomic Fountain Frequency Standard Prototype at NIST. In: *Proc. 1995 Frequency Standard Metrology Symposium*, pp. 411–412, October 1995.
46. L. Essen, J.V.L. Parry. Atomic standard of frequency and time interval. *Nature*, 176, 1955.
47. A.M. Finkelstein, A.V. Ipatov, Z.M. Malkin, E.A. Skurikhina, S.G. Smolentsev. Results of the first two years of VLBI observations at the Svetloe Ob-

servatory within the framework of international geodynamical programs. *Astronomy Letters*, 32:138–144, February 2006.
48. P. Forman. Atomichron: The atomic clock from concept to commercial product.
http://www.ieee-uffc.org/freqcontrol/atomichron/automichron.htm.
49. Y. Fukuzaki, K. Shibuya, K. Doi, T. Ozawa, A. Nothnagel, T. Jike, S. Iwano, D.L. Jauncey, G.D. Nicolson, P.M. McCulloch. Results of the VLBI experiments conducted with Syowa Station, Antarctica. *Journal of Geodesy*, 79:379–388, August 2005.
50. V. Gerginov, S. Knappe, P. Schwindt, V. Shah, L. Hollberg, J. Kitching. Long-term frequency instability of atomic frequency references based on coherent population trapping and microfabricated vapor cells. *Journal of the Optical Society of America B*, 23(4), April 2006.
51. W. Gerlach, O. Stern. Uber die richtungsquantelung im magnetfeld. *Annalen der Physik*, IV, 1924.
52. H.M. Goldenberg, D. Kleppner, N.F. Ramsey. Atomic hydrogen maser. *Physics Review Letters*, 5, 1960.
53. C. Hackman, J. Levine, T.E. Parker, D. Piester, J. Becker. A new technique for estimating frequency from GPS carrier-phase time transfer data. In: *2004 IEEE International Ultrasonics, Ferroelectrics, and Frequency Control Joint 50th Anniversary Conference*, August 2004.
54. D.W. Hanson. Fundamentals of two-way time transfer by satellite. In: *Proc. 1989 IEEE Freq. Cont. Symp.*, May 1989.
55. A. Hasegawa, K. Fukuda, M. Kajita, H. Ito, M. Kumagai, M. Hosokawa, N. Kotake, T. Morikawa. Accuracy evaluation of optically pumped primary frequency standard CRL-O1. *Metrologia*, 41:257–263, May 2004.
56. T.P. Heavner, S.R. Jefferts, E.A. Donley, T.E. Parker, F. Levi. A new microwave synthesis chain for the primary frequency standard NIST-F1. In: *Proc. 2005 Joint Mtg. IEEE Intl. Freq. Cont. Symp. and PTTI*, volume 57, pp. 308–311, August 2005.
57. T.P. Heavner, S.R. Jefferts, E.A. Donley, J.H. Shirley, T.E. Parker. NIST-F1: Recent improvements and accuracy evaluations. *Metrologia*, 42:411–422, 2005.
58. T.P. Heavner, S.R. Jefferts, E.A. Donley, J.H. Shirley, T.E. Parker. Recent improvements in NIST-F1 and a resulting accuracy of $\delta f/f = 0.61 \times 10^{-15}$. *IEEE Transactions on Instrumentation and Measurement*, 54(2):842–845, April 2005.
59. J. Henton, M. Craymer, J. Liard, R. Duval, C. Klatt. Integration of the Canadian gravity standardization network with the Canadian spatial reference system: Challenges and opportunities. *AGU Fall Meeting Abstracts*, pp. B362+, December 2005.
60. I.M. Hoffman, W.M. Goss, P. Palmer. VLBA+Y27 images of the formaldehyde masers in NGC 7538 and G29.96-0.02. In: J. Romney, M. Reid (eds.), *ASP Conf. Ser. 340: Future Directions in High Resolution Astronomy*, pp. 347–+, November 2005.
61. T. Hong, C. Cramer, W. Nagourney, N. Fortson. Prospects for an Optical Frequency Standard in Atomic Yb. *APS Meeting Abstracts*, pp. J1086+, May 2004.
62. D.A. Howe, D.W. Hanson, J.L. Jespersen, M.A. Lombardi. Satellite two-way time transfer: Fundamentals and recent progress. In: *Procedings of the 1989 Precise Time and Time Interval (PTTI) Meeting*, December 1989.

63. Information Technology Laboratory. Engineering statistics handbook. http://www.itl.nist.gov/div898/handbook/index.htm.
64. Information Technology Laboratory. Statistical methods of uncertainty analysis.
http://www.itl.nist.gov/div898/carroll/u.htm.
65. International Astronomical Union. Recommendation resolution B 1.8. In: *Proceedings of the 24th General Assembly, Manchester 2000*, volume XXIV B, 2001.
66. J. Terrien. News from the International Bureau of Weights and Measures. *Metrologia*, 4(1):41–45, 1968.
67. J. Terrien. News from the Bureau International des Poids et Mesures. *Metrologia*, 7(1):43–44, 1971.
68. N. Jakowski, R. Leitinger, L. Ciraolo. Behaviour of large scale structures of the electron content as a key parameter for range errors in GNSS applications. *Annals of Geophysics*, supplement to vol 47(2/3):1031–1047, 2004.
69. S.R. Jefferts, D.M. Meekhof, L. Hollberg, R.E. Drullinger, W.D. Lee, F.L. Walls, C.W. Nelson, F. Levi, T.E. Parker. NIST Cesium fountain frequency standard: Preliminary Results. In: *Proc. 1998 IEEE Intl. Freq. Cont. Symp.*, pp. 2–5, May 1998.
70. M.A. Kasevich, E. Riis, S. Chu, R.G. Devoe. RF spectroscopy in an atomic fountain. *Physical Review Letters*, 63:612–615, 1989.
71. D. Kleppner, H.M. Goldenberg, N.F. Ramsey. Theory of the hydrogen maser. *Physics Review*, 126, 1962.
72. S. Knappe, V. Shah, P. Schwindt, L. Hollberg, J. Kitching. A microfabricated atomic clock. *Applied Physics Letters*, 85(9), August 2004.
73. Y.Y. Kovalev, K.I. Kellermann, M.L. Lister, D.C. Homan, R.C. Vermeulen, M.H. Cohen, E. Ros, M. Kadler, A.P. Lobanov, J.A. Zensus, N.S. Kardashev, L.I. Gurvits, M.F. Aller, and H.D. Aller. Sub-milliarcsecond imaging of quasars and active galactic nuclei. IV. Fine-scale structure. *The Astronomical Journal*, 130:2473–2505, December 2005.
74. T. Kurosu, Y. Fukuyama, Y. Koga, K. Abe. Preliminary evaluation of the Cs atomic fountain frequency standard at NMIJ/AIST. *IEEE Transactions on Instrumentation and Measurement*, 53(2):466–471, April 2004.
75. K. Larson, J. Levine. Time transfer using the phase of the GPS carrier. *IEEE Transactions on Ultrasonics, Ferroelectrics and Frequency Control*, 45(3), May 1998.
76. K. Larson, J. Levine. Carrier-phase time transfer. *IEEE Transactions on Ultrasonics, Ferroelectrics and Frequency Control*, 46(4), July 1999.
77. W.L. Lawrence. Atom time piece displayed here. *New York Times*, October 4 1956.
78. H.S. Lee, T.Y. Kwon, H.S. Kang, Y.H. Park, C.H. Oh, S.E. Park, H. Cho, V.G. Minogin. Comparison of the Rabi and Ramsey pulling in an optically pumped caesium-beam standard. *Metrologia*, 40:224–231, October 2003.
79. P. Lesage, C. Audoin. Characterization of frequency stability: Uncertainty due to the finite number of measurements. *IEEE Transactions on Instrumentation and Measurement*, 22(2), 1973.
80. P. Lesage, C. Audoin. Estimation of the two-sample variance with a limited number of data. In: *31st Annual Symposium on Frequency Control, 1977*, 1977.
81. P. Lesage, C. Audoin. Characterization and measurement of time and frequency stability. *Radio Science*, 14(4):521–539, July–August 1979.

82. P.D. Lett, W.D. Phillips, S.L. Rolston, C.E. Tanner, R.N. Watts, C.I. Westbrook. Optical molasses. *Optical Society of America Journal B Optical Physics*, 6:2084–2107, 1989.
83. F. Levi, L. Lorini, D. Calonico, A. Godone. Systematic shift uncertainty evaluation of IEN CSF1 primary frequency standard. *IEEE Transactions on Instrumentation and Measurement*, 52(2):267–271, April 2003.
84. L. Lo Presti, D. Rovera, A. De Marchi. A simple analysis of the Dick effect in terms of phase noise spectral densities. *IEEE Transactions on Ultrasonics, Ferroelectrics and Frequency Control*, 45(4), 1998.
85. M.A. Lombardi, L.M. Nelson, A.N. Novick, V.S. Zhang. Time and frequency measurements using global positioning system. *Cal. Lab. Int. J. Metrology*, IM-34, July–September 2001.
86. H. Lyons. The atomic clock. *Instruments*, 22, 1949.
87. H. Lyons. The atomic clock. *American Scholar*, 19, April 1950.
88. C. Ma, D.S. MacMillan. The role of VLBI in the TRF. *AGU Fall Meeting Abstracts*, pp. A3+, December 2005.
89. E. Majorana. Atomi orientati in campo magnetico variabile. *Nuovo Cimento*, 9:43–50, 1932.
90. A. Makdissi, E. de Clercq. Evaluation of the accuracy of the optically pumped caesium beam primary frequency standard of BNM-LPTF. *Metrologia*, 38:409–425, 2001.
91. A. Makdissi, E de Clercq. Evaluation of the accuracy of the optically pumped caesium beam primary frequency standard of BNM-LPTF. *Metrologia*, 38:409–425, October 2001.
92. A.G. Mann, C. Sheng, A.N. Luiten. Cryogenic sapphire oscilator with exceptionally high frequency stability. *IEEE Transactions on Instrumentation and Measurement*, 50(2), April 2001.
93. A.A. Michelson, F.G. Pease. Measurement of the diameter of alpha Orionis with the interferometer. *Astrophysical Journal*, 53:249–259, May 1921.
94. P.J. Mohr, B.N. Taylor. CODATA recommended values of the fundamental physical constants. *Reviews of Modern Physics*, 77:1–107, 2002.
95. J. Nastula, B. Kolaczek, R. Weber, J. Boehm, H. Schuh. High resolution Earth rotation parameters determined during the CONT02 campaign. In: N.R. Vandenberg, K.D. Baver (eds.) *International VLBI Service for Geodesy and Astrometry 2004 General Meeting Proceedings*, 2004.
96. NIST. Time and Frequency from A to Z. http://tf.nist.gov/general/glossary.htm.
97. C. Oates, E.A. Curtis, G. Wilpers, C. Hoyt, L. Hollberg, A. Ludlow. A Ca optical frequency standard based on colder atoms. *APS Meeting Abstracts*, pp. B4005+, May 2003.
98. Oscilloquartz S.A. *OCXO 8600 10 times better than any other VCXO*. Oscilloquartz S.A.
99. W.H. Oskay, T. Rosenband, U. Tanaka, C.E. Tanner, S.A. Diddams, L. Hollberg, W.M. Itano, J.C. Bergquist. The mercury ion optical clock. *APS Meeting Abstracts*, pp. J1050+, May 2004.
100. P. Giacomo. News from the BIPM. *Metrologia*, 17(1):69, 1981.
101. T.E. Parker, S.R. Jefferts, T.P. Heavner, E.A. Donley. Operation of the NIST-F1 cesium fountain primary frequency standard with a maser ensemble, including the impact of frequency transfer noise. *Metrologia*, 42:423–430, 2005.

102. T.E. Parker, V. Zhang. Sources of instabilities in two-way satellite time transfer. In: *Proceedings of the 2005 Joint Meeting. IEEE International Frequency Control Symposium and PTTI*, August 2005.
103. M.A. Pavlis, N.K. Weiss. The relativistic redshift with $3 \times 10-17$ uncertainty at NIST, Boulder, Colorado, USA. *Metrologia*, 40(2):66–73, 2003.
104. W.D. Phillips. Laser cooling and trapping of neutral atoms. *NASA STI/Recon Technical Report N*, 82:14512–+, 1981.
105. Precise Time and Frequency, Inc. *CH1-75 Active Hydrogen Maser.* See for instance http://www.ptfinc.com/images/CH1-75%20Intro.pdf.
106. T.J. Quinn. Base units of the système international d'unités, their accuracy, dissemination and international traceability. *Metrologia*, 31:515–527, January 1995.
107. N.F. Ramsey. Molecular beam resonance method with separated oscillating fields. *Physics Review*, 78, 1950.
108. J.D. Romney. The VLBA correlator. *Bulletin of the American Astronomical Society*, 27, May 1995.
109. M. Ryle, D. Vonberg. Solar radiation on 175Mc/s. *Nature*, 158, September 1946.
110. G. Sagnac. Practical implications of relativity for a global coordinate time scale. *C.R. Acad. Sci. Paris*, 157:708–710, 1410–1413, 1913.
111. W. Schlueter, E. Himwich, A. Nothnagel, N. Vandenberg, A. Whitney. IVS and its important role in the maintenance of the global reference systems. *Advances in Space Research*, 30(2), July 2002.
112. Schriever AFB. Constellation status. http://www.schriever.af.mil/gps/.
113. J.H. Shirley, W.D. Lee, R.E. Drullinger. Accuracy evaluation of the primary frequency standard NIST-7. *Metrologia*, 38:427–458, 2001.
114. Symmetricom. *5071A Primary Frequency Standard.* Symmetricom.
115. C. Tamm, T. Schneider, E. Peik. Optical frequency standard based on a trapped $^{171}Yb^{+}$ ion. In: G. Huber, I.A. Scherbakov, V.Y. Panchenko (eds.) *International Conference on Lasers, Applications, and Technologies 2002: Advanced Lasers and Systems. Edited by Huber, Guenter; Scherbakov, Ivan A., Panchenko, Vladislav Y. Proceedings of the SPIE, Volume 5137, pp. 403–407 (2003).*, pp. 403–407, 2003.
116. U. Tanaka, S. Bize, C.E. Tanner, R.E. Drullinger, S.A. Diddams, L. Hollberg, W.M. Itano, D.J. Wineland, J.C. Bergquist. The $^{199}Hg^{+}$ single ion optical clock: recent progress. *Journal of Physics B Atomic Molecular Physics*, 36:545–551, 2003.
117. R. Tausworthe. Convergence of oscillator spectral estimators for counted-frequency measurements. *IEEE Transactions on Communications*, 20(2), April 1972.
118. M. Trott. *The Mathematica GuideBook for Programming.* Springer, Berlin Heidelberg New York, 2004.
119. J.S. Ulvestad, K.E. Johnson, S.G. Neff. A VLBI search for radio supernovae in super star clusters. *American Astronomical Society Meeting Abstracts*, 207, December 2005.
120. US Coast Guard Navigation Center. Global Positioning System Standard Positioning Service Signal Specification. http://www.navcen.uscg.gov/pubs/gps/sigspec/gpssps1.pdf.

121. US Coast Guard Navigation Center. Global Positioning System Standard Positioning Service Signal Specification – Annex A thru C. http://www.navcen.uscg.gov/pubs/gps/sigspec/gpsspsa.pdf.
122. USNO. Block II satellite information. ftp://tycho.usno.navy.mil/pub/gps/gpsb2.txt.
123. USNO. GPS constellation status. ftp://tycho.usno.navy.mil/pub/gps/gpstd.txt.
124. J. Vanier, C. Audoin. *The Quantum Physics of Atomic Frequency Standards.* Hilger, Bristol.
125. C. Vian, P. Rosenbusch, H. Marion, S. Bize, L. Cacciapuoti, S. Zhang, M. Abgrall, D. Chambon, I. Maksimovic, P. Laurent, G. Santarelli, A. Clairon, A. Luiten, M. Tobar, C. Salomon. BNM-SYRTE fountains: Recent results. *IEEE Transactions on Instrumentation and Measurement*, 54(2):833–836, April 2005.
126. C. Vian, P. Rosenbusch, H. Marion, F. Pereira Dos Santos, M. Abgrall, S. Zhang, Y. Sortais, S. Bize, I. Maksimovic, D. Calonico, J. Grünert, C. Mandache, P. Lemonde, G. Santarelli, Ph. Laurent, A. Clairon. Recent results of BNM-SYRTE fountain clocks. Joint Meeting IEEE International Frequency Control Symposium and 17th European Frequency and Time Forum, May 2003.
127. W. Fricke, H. Schwan, T. Lederle, U. Bastian, R. Bien, G. Burkhardt, B. du Mont, R. Hering, R. Jährling, H. Jahreiß, S. Röser, H.-M. Schwerdtfeger, H.G. Walter. Fifth fundamental catalogue (FK5). part 1. the basic fundamental stars. *Veröffentlichungen Astronomisches Rechnen-Institut Heidelberg*, (32), 1988.
128. E.W. Weisstein. Ergodic theory. From MathWorld–A Wolfram Web Resource. http://mathworld.wolfram.com/ErgodicTheory.html.
129. E.W. Weisstein. Random walk. From MathWorld–A Wolfram Web Resource. http://mathworld.wolfram.com/RandomWalk.html.
130. E.W. Weisstein. Random walk–1-dimensional. From MathWorld–A Wolfram Web Resource. http://mathworld.wolfram.com/RandomWalk1-Dimensional.html.
131. Wikipedia the Free Encyclopedia. Hadamard transform. http://en.wikipedia.org/wiki/Hadamard_transform.
132. Wikipedia the Free Encyclopedia. Nyquist–shannon sampling theorem. http://en.wikipedia.org/wiki/Nyquist–Shannon_sampling_theorem.
133. Wikipedia the Free Encyclopedia. Pink noise. http://en.wikipedia.org/wiki/Flicker_noise.
134. Wikipedia the Free Encyclopedia. Stark effect. http://en.wikipedia.org/wiki/Stark_effect.
135. Wikipedia the Free Encyclopedia. Stationary process. http://en.wikipedia.org/wiki/Stationary_process.
136. Wikipedia the Free Encyclopedia. White noise. http://en.wikipedia.org/wiki/White_noise.
137. Wikipedia the Free Encyclopedia. Zeeman effect. http://en.wikipedia.org/wiki/Zeeman_effect.
138. G.M.R. Winkler. Modified julian date. http://tycho.usno.navy.mil/mjd.html.

139. K. Yoshimura. Characterization of frequency stability: Uncertainty due to the autocorrelation of the frequency fluctuations. *IEEE Transactions on Instrumentation and Measurement*, 27(1), 1973.
140. P. Zeeman. The Effect of magnetisation on the nature of light emitted by a substance. *Nature*, 55, February 1897.

Index

Accuracy
  Definition, 57
  Typical values, 60
    Commercial devices, 61
    Primary frequency standards, 60
Aliasing, 72
Allan filter, 93
Allan variance
  Definition, 93
  Flicker frequency noise, 97
  Flicker phase noise, 97
  Frequency drift, 97
  Frequency response, 94
  Random walk frequency noise, 97
  Uncertainty, 99
  White frequency noise, 96
  White phase noise, 97
Atom
  Fine structure, 15
  Hyperfine structure, 15
Atomic clock
  Modelization, 137
Atomic fountain
  Atomic signal amplitude noise frequency noise, 51
  OL frequency noise, 49
  Preparation of the atoms, 31
  principle, 30
  Probing of the transition, 31
  Servo-loop, 48

Barycentric celestial reference system (BCRS), 135
Barycentric coordinate time (TCB), 135

CCDS, 131
CCTF, 131
Cesium atom
  Frequency of the hyperfine transition, 18
Cesium frequency standard
  Amplitude noise of the atomic signal, 44
  Atomic beam preparation, 20
  Detection of the transition, 23
  Excitation of the transition, 20
  Frequency error, 46
  Frequency noise of the atomic signal, 43
  Generation of the probing signal, 25
  OL frequency noise, 43
  Ramsey cavity, 20
  Servo control of the local oscillator, 27
  Transition probability, 20
Circular T, 133
Clock transition, 15
Coordinated universal time, 136
Covariance, 75

Definition of the second (1967), 18
Dick effect, 32

EAL – échelle atomique libre, 132
Ergodicity, 76

Flicker frequency noise, 84
Flicker phase noise, 84
Frequency fluctuations
  components, 65
Frequency noise vs phase noise, 83
Frequency standard
  Active and passive, 16
  Passive
    Optically pumped cesium beam, 16

Galileo, 173
  Control segment, 173
  Data, 174
  Space segment, 173
Geocentric celestial reference system (GCRS), 135
Geocentric coordinate time (TCG), 135
GLONAS, 172
  Control segment, 172
  Data, 173
  Space segment, 172
GPS
  Algorithm for a receiver, 200
  Anti-spoofing, 156
  C/A-code, 155
  C/A-code ambiguity, 162
  Carrier L_1, 155
  Carrier L_2, 155
  Carrier phase tracking, 168
  Coarse acquisition code, 155
  Control segment, 153
  Data emitted by the satellites, 154
  Data processing by the receiver, 161
  Differential GPS, 167
  Doppler measurement, 171
  EGNOS, 172
  Error sources, 166
  Geometric dilution of precision (GDOP), 165
  GPS time, 160
  LAAS, 172
  MSAS, 172
  Navigation message, 158, 162
  P-code, 155
  P-code ambiguity, 162
  P-code processing, 167
  Precise code, 155
  Propagation delays, 163
  Quick description, 151
  Satellite time, 159
  Selective availability, 156
  Space segment, 152
  Spread spectrum technique, 154
  User segment, 154
  WAAS, 171
GPS time scale, 136
GPS time scale vs. UTC, 137

Hadamard filter, 100
Hadamard variance
  Definition, 99, 100
  Flicker frequency noise, 105
  Flicker phase noise, 104
  Frequency drift, 105
  Random walk frequency noise, 105
  White frequency noise, 104
  White phase noise, 105
Hydrogen atom
  Hyperfine transition frequency, 28
Hydrogen maser
  Atomic beam preparation, 28
  Atomic signal frequency noise, 37
  Electronic package, 30
  Frequency error, 38
  Hyperfine frequency emission, 28
  LO frequency noise, 36
  Maser effect, 28
  Output phase noise, 38
  Phase control loop, 34
  Phase locking of the local oscillator, 29

IAU, 134
Ionospheric delay, 144

Leap seconds, 136
  Since 1972, 136
Localization system, 151

Nyquist–Shannon sampling theorem, 73

Oscillator
  Atomic resonance, 15
  Electrical resonance, 14
  Mechanical resonance, 14
  The two parts, 13

Ramsey fringes, 21
Random walk frequency noise, 83

Sagnac effect, 143, 144, 149, 150
SI second
  Definition, 18
SI units, 129
Sidereal time, 134
Spectral density, 76
  and Fourier transform, 77
  and variance, 77
  filtering, 80
Stability

Definition, 65
Frequency measurement, 67
Measurement in time domain: principle, 67
Stationarity
First-order, 75
Second-order, 75
Synchronization
Carrier phase GPS, 146
Einstein clocks synchronization, 142
GPS common-view, 145
One-way GPS, 143
Rotating frame, 143
Two-way satellite, 147
Synchronization of Einstein clocks, 142

TAI
Accuracy, 132
Definition, 131
Processing, 132
Temps atomique international, 131
Processing, 132
Three-samples variance, 110
Flicker frequency noise, 111
Flicker phase noise, 111
Random walk frequency noise, 112
White frequency noise, 111
White phase noise, 111
Time scale
Definition, 130
Requirements, 130
Time scales
Algorithms, 140
Weight of the clocks, 142
Tropospheric delay, 144
TWSTFT, 147

Uncertainty
Definition, 57
Example of evaluation, 58
Principle of evaluation, 58
Type A, 58
Type B, 58
Typical values, 60
Commercial devices, 61
Primary frequency standards, 60
Uncertainty of the SI base units, 129
Universal time, 3, 134
UT, 3, 134
UT1, 134
UTC, 136
UTC vs. UTC(k), 137

Variance, 75
Discrete-time vs. continuous-time, 80
Unbiased estimator, 85
Variances
Analysis of the frequency noise, 113
Frequency drift, 119
Sensibility to the noise components, 115
Summary, 112
Summary of the bandwidths, 112
Very long base interferometry, 175
VLBI, 175
Coherence time, 181
Correlator, 186
Cross-correlation, 181
Earth rotation, 188
Fourier transform, 183
Fringe visibility, 181
Imaging, 183, 187
International Terrestrial Reference System, 188
Local oscillator, 186
Plate motions, 188
Principle, 176
Resolution, 176
Timestamp, 186

White frequency noise, 84
White phase noise, 84

Printing: Krips bv, Meppel
Binding: Stürtz, Würzburg